T0225939

Lecture Notes in Mathematics

Edited by A. Dold, F. Takens and B. Teissier

Editorial Policy
for the publication of monographs

1. Lecture Notes aim to report new developments in all areas of mathematics – quickly, informally and at a high level. Monograph manuscripts should be reasonably self-contained and rounded off. Thus they may, and often will, present not only results of the author but also related work by other people. They may be based on specialized lecture courses. Furthermore, the manuscripts should provide sufficient motivation, examples and applications. This clearly distinguishes Lecture Notes from journal articles or technical reports which normally are very concise. Articles intended for a journal but too long to be accepted by most journals, usually do not have this "lecture notes" character. For similar reasons it is unusual for doctoral theses to be accepted for the Lecture Notes series.

2. Manuscripts should be submitted (preferably in duplicate) either to one of the series editors or to Springer-Verlag, Heidelberg. In general, manuscripts will be sent out to 2 external referees for evaluation. If a decision cannot yet be reached on the basis of the first 2 reports, further referees may be contacted: the author will be informed of this. A final decision to publish can be made only on the basis of the complete manuscript, however a refereeing process leading to a preliminary decision can be based on a pre-final or incomplete manuscript. The strict minimum amount of material that will be considered should include a detailed outline describing the planned contents of each chapter, a bibliography and several sample chapters.
Authors should be aware that incomplete or insufficiently close to final manuscripts almost always result in longer refereeing times and nevertheless unclear referees' recommendations, making further refereeing of a final draft necessary.
Authors should also be aware that parallel submission of their manuscript to another publisher while under consideration for LNM will in general lead to immediate rejection.

3. Manuscripts should in general be submitted in English.
Final manuscripts should contain at least 100 pages of mathematical text and should include
– a table of contents;
– an informative introduction, with adequate motivation and perhaps some historical remarks: it should be accessible to a reader not intimately familiar with the topic treated;
– a subject index: as a rule this is genuinely helpful for the reader.

Lecture Notes in Mathematics 1731

Springer
Berlin
Heidelberg
New York
Barcelona
Hong Kong
London
Milan
Paris
Singapore
Tokyo

Tim Hsu

Quilts:
Central Extensions,
Braid Actions, and
Finite Groups

 Springer

Author

Tim Hsu
Department of Mathematics
Pomona College
610 N. College Ave.
Claremont
CA 91711-6348, USA

E-mail: timhsu@pccs.cs.pomona.edu

Cataloging-in-Publication Data applied for

Die Deutsche Bibliothek - CIP-Einheitsaufnahme

Hsu, Tim:
Quilts : central extensions, braid actions, and finite groups / Tim
Hsu. - Berlin ; Heidelberg ; New York ; Barcelona ; Hong Kong ; London
; Milan ; Paris ; Singapore ; Tokyo : Springer, 2000
 (Lecture notes in mathematics ; 1731)
 ISBN 3-540-67397-0

Mathematics Subject Classification (2000): 20D08, 20E22, 20F36, 20H10

ISSN 0075-8434
ISBN 3-540-67397-0 Springer-Verlag Berlin Heidelberg New York

Springer-Verlag is a company in the BertelsmannSpringer publishing group.
© Springer-Verlag Berlin Heidelberg 2000
Printed in Germany

Typesetting: Camera-ready $T_{E}X$ output by the author
Printed on acid-free paper   SPIN: 10724981   41/3143/du   543210

Preface

Quilts, as developed by Norton, Parker, Conway, and the author, are 2-complexes used to analyze actions and subgroups of the 3-string braid group, or more generally, to analyze actions and subgroups of certain central extensions of triangle groups. This monograph describes the fundamentals of quilt theory, including basic definitions, main results, and applications to central extensions, braid actions, and finite groups. Though the monograph contains some purely expository material, it is primarily a research monograph. Indeed, most of the results have not previously been published in a widely available form, and many results appear here in print for the first time.

The material in this monograph should be accessible to graduate students. More specifically, the only background necessary is a first graduate course in algebra, though some knowledge of algebraic topology and familiarity with finite group theory is helpful. The methods and results may be relevant to researchers interested in finite groups, moonshine, central extensions, triangle groups, dessins d'enfants, and monodromy actions of braid groups.

I would like to take this opportunity to thank Igor Dolgachev, Martin Edjvet, Andy Mayer, Tilman Schulz, Shahriar Shahriari, Dani Wise, and the editorial staff at Springer-Verlag, for their help with this monograph; Adam Deaton, for his program lafite, which was used to draw Fig. 3.5.1; Peter Doyle, for the name "quilts"; John H. Conway, for his help and mentorship; and Simon Norton, both for originating the subject of quilts, and also for his considerable help with this and other work. More personally, I would like to thank my family and friends for their support, encouragement, and patience.

This is dedicated to my parents.

Claremont, CA, February 2000   *Tim Hsu*

Contents

Part II. The structure problem

Appendix

Symbols

1. Introduction

We discuss some of the history and motivation behind quilts and Norton systems and provide an overview of the contents of this monograph.

1.1 Motivation

The principal motivation for this monograph comes from two areas of mathematics: moonshine and central extensions.

The first area, moonshine, ties together finite groups, number theory, Lie algebras, and mathematical physics, and has been the source of much remarkable work over the last twenty years. For a general overview of moonshine, see Borcherds [8] and Goddard [33], on which we rely for much of the following; see also Dong and Mason [27] and Ferrar and Harada [31].

Recall that a discrete subgroup Γ of $\mathbf{PSL}_2(\mathbf{R})$ is said to have *genus zero* if $\Gamma\backslash\mathbf{H}^2$ is a surface of genus zero, where $\mathbf{H}^2$ is the complex upper half-plane. Recall also that a meromorphic function $j : \mathbf{H}^2 \to \mathbf{C}$ is said to be a *Hauptmodul* for a genus zero subgroup Γ of $\mathbf{PSL}_2(\mathbf{R})$ if every Γ-invariant meromorphic function $f : \mathbf{H}^2 \to \mathbf{C}$ is a rational function of j. (See Schoeneberg [76] or Serre [80] for the basic facts on Hauptmoduls and other modular functions.) Finally, recall that the largest sporadic finite simple group is called the Fischer-Griess *Monster*.

Starting from observations of McKay and Thompson, Conway and Norton [18] discovered a phenomenon called *Monstrous Moonshine* that connects Hauptmoduls and the Monster in a deep and surprising fashion. Specifically, Conway and Norton conjectured and provided numerical evidence that there exists an infinite-dimensional representation $V = \bigoplus_{n \in \mathbf{Z}} V_n$ of the Monster such that if H_n is the character of the representation V_n, then for any fixed element g of the Monster, the function

$$T_g(z) = \sum_{n \in \mathbf{Z}} H_n(g)q^n, \tag{1.1.1}$$

where $q = e^{2\pi i z}$, is a Hauptmodul for some genus zero subgroup Γ of $\mathbf{PSL}_2(\mathbf{R})$.

For example, for $g = 1$, $H_n(g)$ takes the values

$$H_{-1}(1) = 1,$$
$$H_1(1) = 196884 \quad = 1 + 196883,$$
$$H_2(1) = 21493760 = 1 + 196883 + 21296876, \qquad (1.1.2)$$

$$\vdots \qquad \vdots$$

where 1, 196883, and 21296876 are the degrees of the smallest irreducible representations of the Monster, and the elliptic modular function

$$T_1(z) = j(z) = q^{-1} + 196884q + 21493760q^2 + \dots \qquad (1.1.3)$$

is a Hauptmodul for $\mathbf{PSL}_2(\mathbf{Z})$.

The fact that every element of the Monster corresponds naturally (via the function $T_g(z)$) to a subgroup of genus zero is one of the more remarkable aspects of Monstrous Moonshine. While this *genus zero phenomenon* has been explained to a large extent by Borcherds' proof of the original conjectures [7] and recent work of Cummins and Gannon [23], the phenomenon is still of interest. More relevantly to the matters at hand, the genus zero phenomenon provided the initial motivation for Norton's invention [62] of the diagrams that we now call quilts (see Sect. 1.2).

While quilts were invented to study moonshine, they are also useful for studying triangle groups and their central extensions. The best-known antecedent for this practice is the use of diagrams of various types to study subgroups of the classical modular group $\mathbf{PSL}_2(\mathbf{Z})$. See Atkin and Swinnerton-Dyer [3, §3], Jones [47, §2], Jones and Singerman [48], Millington [59], and Schneps [73] for examples.

In terms of central extensions of triangle groups, the closest-related previous work is Conway, Coxeter, and Shephard [14], which includes a result giving the order of the group

$$\left\langle a, b, c, z \;\middle|\; \begin{array}{l} 1 = [z, a] = [z, b] = [z, c], \\ a^\ell = z^p, \; b^m = z^q, \; c^n = z^r, \; abc = z^s \end{array} \right\rangle \qquad (1.1.4)$$

when the triangle group

$$\langle a, b, c \mid 1 = a^\ell = b^m = c^n = abc \rangle \qquad (1.1.5)$$

is finite. (See Sect. 2.2 for a brief introduction to triangle groups.) The most interesting feature of this result is that "horizontal" information about the order of the triangle group (1.1.5) gives "vertical" information about the order of z in its central extension (1.1.4). One of the main goals of this monograph is to obtain similar (and related) results for subgroups of triangle groups. See Chaps. 3, 6, and 7 for examples.

For the reader familiar with low-dimensional topology, we note that the central extensions we study also arise naturally as the fundamental groups of *Seifert fibered 3-manifolds*. We therefore call a triangle group extension like (1.1.4) a *Seifert group*. For more on Seifert fibered 3-manifolds and their fundamental groups, see Orlik [69], Scott [78, §1–3], and Seifert [79].

1.2 A brief history of quilts

Quilts and Norton systems have a somewhat involved history that is worth discussing briefly, both as motivation and as an explanation of how these ideas began. We follow Norton [62] for much of our initial exposition.

Let G be a finite group, let L, resp. R, be the element $\begin{pmatrix} 1 & 1 \\ 0 & 1 \end{pmatrix}$, resp. $\begin{pmatrix} 1 & 0 \\ 1 & 1 \end{pmatrix}$, of $\mathbf{SL}_2(\mathbf{Z})$, and let P be the set of all pairs of commuting elements of G. $\mathbf{SL}_2(\mathbf{Z})$ acts on P by

$$(g, h) \begin{pmatrix} a & b \\ c & d \end{pmatrix} \mapsto (g^a h^c, g^b h^d). \tag{1.2.1}$$

In particular, for $i \in \mathbf{Z}$, we have

$$(g, h)L^i = (g, g^i h),$$
$$(g, h)R^i = (gh^i, h). \tag{1.2.2}$$

We can therefore say that a function $F : P \times \mathbf{H}^2 \to \mathbf{C}$ is *automorphic* with respect to $\mathbf{SL}_2(\mathbf{Z})$ if

$$F((g^a h^c, g^b h^d), z) = F\left((g, h), \frac{az + b}{cz + d}\right) \tag{1.2.3}$$

for all $\begin{pmatrix} a & b \\ c & d \end{pmatrix} \in \mathbf{SL}_2(\mathbf{Z})$. For instance, when G is trivial, and $F((g, h), z)$ is meromorphic for $z \in \mathbf{H}^2$, $f(z) = F((1, 1), z)$ is an ordinary modular function over $\mathbf{SL}_2(\mathbf{Z})$. For a general commuting pair (g, h) and meromorphic F, $f(z) = F((g, h), z)$ is modular over some congruence subgroup of $\mathbf{SL}_2(\mathbf{Z})$.

The above construction turns out to provide an interesting framework for studying moonshine, especially in terms of conformal field theory; see, for instance, Mason [56] or Tuite [85]. However, Norton was just as interested in *non*-commuting pairs (g, h), for even then, it can be shown that the rule in (1.2.2) still defines an action of $\mathbf{SL}_2(\mathbf{Z})$ on conjugacy classes of pairs of elements of G. An orbit of this action is the forerunner of what we now call a *Norton system*.

One interesting feature of extending the action (1.2.2) to non-commuting pairs is that we can still use (1.2.3) to define a notion of automorphic function, as long as we add the condition

$$F((g^\tau, h^\tau), z) = F((g, h), z) \tag{1.2.4}$$

for all $\tau \in G$. This provides a relatively natural connection between automorphic functions and finite groups, and in fact, between genus 0 modular subgroups and the Monster, as we shall see later (Sect. 11.1). Unfortunately, the investigation of such automorphic functions is beyond the scope of this monograph. Instead, our main goal in this area is to understand the structure of Norton systems.

The key to understanding the structure of Norton systems is to represent them by a specialized type of coset diagram. Norton initially called these diagrams *footballs*, and they are mentioned in [62] as "trivalent polyhedra". (See Chap. 11 for an explanation of the name "football".) From these origins, footballs were further developed and refined by Norton, Parker, and Conway. The most important refinement of the basic idea came with considering the elements L and R in (1.2.2) to be elements of $\mathbf{B}_3$, the 3-string braid group, and using (1.2.2) to lift the $\mathbf{SL}_2(\mathbf{Z})$ action to an action of $\mathbf{B}_3$ on pairs of elements in G. This $\mathbf{B}_3$-action, which we call *Norton's action*, may then be represented by appropriately decorated diagrams, which are what we now call *quilts*. (Note that Norton has recently proposed the name "nets" for a certain class of quilts; see Sect. 13.1 for details.)

Later, the first published formal definitions of quilts and Norton systems (then called *T-systems*) appeared in Conway and Hsu [17], along with the first published rigorous proofs of some of their important properties. In this monograph, we give a new account of the fundamentals of quilt theory, superseding both [17] and [43], and proceed from there to obtain many other results. First, however, to make matters a bit more concrete, we have the following intuitive description of how to draw the quilt of a Norton system.

1.3 Drawing the quilt of a Norton system

Suppose we have a pair of elements (α_0, β_0) of a group G, and we wish to understand the structure of $N(\alpha_0, \beta_0)$, the orbit of (α_0, β_0) under Norton's action. In other words, we wish to determine the isomorphism class of $N(\alpha_0, \beta_0)$ as a transitive permutation representation of $\mathbf{B}_3$. (See Chaps. 3 and 4 for more precise statements.) As mentioned previously, $N(\alpha_0, \beta_0)$ is called the *Norton system* of (α_0, β_0), and the picture we will draw to represent it is the *quilt* of $N(\alpha_0, \beta_0)$. As we will see in Chap. 4, in practice, the following rules are often enough to determine the quilt of $N(\alpha_0, \beta_0)$.

Fig. 1.3.1. Typical seam   **Fig. 1.3.2.** First seam

1. The basic unit of a quilt is the *seam*. A typical seam is shown in Fig. 1.3.1. The object at the top of Fig. 1.3.1 is called the *dash* of the seam,

and the object at the bottom is called the *dot*. Each side of a seam is labelled with an element of G, and the line in the middle of a seam is labelled with a number of directed arrows, possibly 0 (as is the case in Fig. 1.3.1).

2. Either 2 or 1 seams meet at a dash, in what is called an *edge* or *collapsed edge*, respectively. Similarly, either 3 or 1 seams meet at a dot, in what is called a *vertex* or *collapsed vertex*, respectively.

3. The first step in drawing the quilt of the Norton system of (α_0, β_0) is to draw the seam in Fig. 1.3.2.

4. The *edge rule* says that if we have already drawn the seam on the left hand side of Fig. 1.3.3, we can complete it to the "directed edge" on the right hand side. Note the arrow on the new (added) seam.

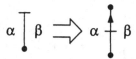

Fig. 1.3.3. Edge rule

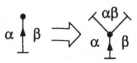

Fig. 1.3.4. Vertex rule

5. The *vertex rule* says that if we have already drawn the seam shown on the left hand side of Fig. 1.3.4, labelled with exactly 1 arrow as indicated, we can complete it to the vertex on the right hand side. Note that both new seams are labelled with 0 arrows.

6. A typical *patch* is shown in Fig. 1.3.5. The *patch rule* says that if α has order n, then the patch of α has n sides. The case shown in Fig. 1.3.5 is $n = 4$. Note that we have not drawn any arrows in Fig. 1.3.5; this is because we cannot immediately determine the number of arrows on a seam that has been added because of the patch rule.

Rules (1) and (2) will later motivate the definitions in Sect. 3.2, and rules (3)–(6) will become Prop. 4.3.1, Prop. 4.3.2, Cor. 4.3.5, and Cor. 4.3.7, respectively.

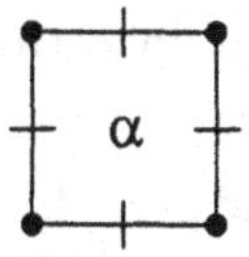

Fig. 1.3.5. Patch rule

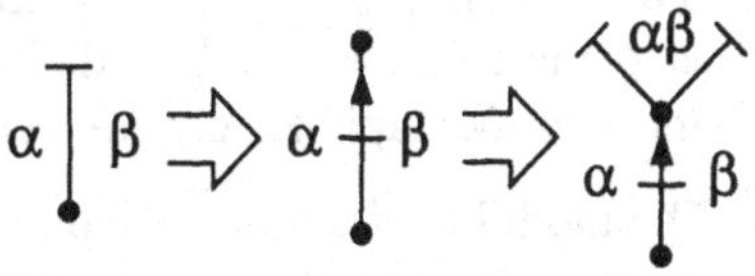

$$\alpha \mathbin{\big|} \beta \Rightarrow \alpha \mathbin{+} \beta \Rightarrow \underset{\alpha \ \ \beta}{\langle \alpha\beta \rangle}$$

Fig. 1.3.6. First steps in Exmp. 1.3.1

Example 1.3.1. Let $G = A_5$, $\alpha = (0\ 1)(2\ 3)$, and $\beta = (0\ 2\ 4)$. Applying rules (3), (4), and (5), we obtain the sequence in Fig. 1.3.6. Note that the action of L (resp. R) on our starting pair (α, β) is represented by a left (resp. right) turn in the diagram. (Compare (1.2.2).)

Next, we compute $\alpha\beta = (0\ 1\ 2\ 3\ 4)$. Applying rule (6), we see that the patch of α is a 2-gon, the patch of β is a 3-gon, and the patch of $\alpha\beta$ is a 5-gon, yielding the first diagram in Fig. 1.3.7. Then, applying rule (2) repeatedly, we identify the remaining open sides of these three patches, as shown in Fig. 1.3.7. We have colored in the patches in the first two steps of Fig. 1.3.7 to indicate how these side identifications force us to close up the diagram to form a sphere. In particular, the closed-up diagram at the end of Fig. 1.3.7 is drawn on a sphere, with the patch of $\alpha\beta$ including the point at infinity.

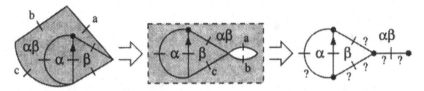

Fig. 1.3.7. Applying rules (6) and (2)

Now, as we will see later, our results so far do not yet completely describe Norton's action on pairs of elements of G. For example, in Fig. 1.3.7, the problem is that the number of arrows on each of the seams labelled with question marks is yet to be determined. Therefore, to describe Norton's action completely, we need the following additional rules. The metaphor to keep in mind is that the arrows represent a flow on the diagram.

7. Every quilt has an associated nonnegative integer M, which we call the *modulus* of the quilt, and the number of arrows on a seam of the quilt is always considered as an integer mod M.
8. The flow out of an edge is 1 mod M, and the flow out of a collapsed edge is $\frac{1}{2}$ mod M, where $\frac{1}{2}$ is defined to be the multiplicative inverse of 2 mod M. In terms of our guiding metaphor, edges are therefore "sources".
9. The flow into a vertex is 1 mod M, and the flow into a collapsed vertex is $\frac{1}{3}$ mod M, where $\frac{1}{3}$ is defined to be the multiplicative inverse of 3 mod M. Metaphorically, vertices are "sinks".

The modulus of a quilt is defined in Sect. 3.3, and Sect. 4.2 describes how to compute the modulus of the quilt of a Norton system. As for rules (8) and (9), they will become definitions in Sect. 3.3.

Example 1.3.2. It turns out that the modulus of the quilt in Exmp. 1.3.1 is 5. Applying rules (8) and (9) to the last diagram in Fig. 1.3.7, and noting that $2 \equiv \frac{1}{3}$ (mod 5), we obtain the diagram in Fig. 1.3.8. In fact, even if we

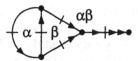

Fig. 1.3.8. Final result

do not know M, rules (8) and (9) allow us to deduce that all of the arrow flows are as shown, which *implies* that M must divide 5, since we must have $2 \equiv \frac{1}{3}$ (mod M). As we will see, this is really an example of the "horizontal implies vertical" results we will obtain in Chap. 6.

1.4 Overview

This monograph has two main parts. Part I (Chaps. 3–7) concentrates on the basic definitions and fundamental results on quilts and Norton systems, and Part II (Chaps. 8–13) addresses the *structure problem* (briefly described below), especially as it relates to finite groups.

In Chap. 2, we review some background material. Of particular note are the sections on triangle groups (Sect. 2.2), *Seifert groups* (Sect. 2.3), and braid groups (Sect. 2.4), since they contain non-standard notation used frequently in the rest of the monograph.

Part I of the monograph begins with Chap. 3. There, we first define modular quilts to be certain 2-complexes and show that modular quilts classify subgroups of triangle groups (or alternately, transitive permutation representations of triangle groups). We then define quilts to be certain equivalence classes of annotated modular quilts, and show that quilts classify subgroups of Seifert groups (or alternately, transitive permutation representations of Seifert groups). The main new idea is that, since Seifert groups are central extensions of triangle groups, any central information that we lose by considering a subgroup of a Seifert group as a subgroup of a triangle group may be regained by a suitable annotation.

Next, in Chap. 4, we define *Norton systems* to be the orbits of *Norton's action* of $\mathbf{B}_3$ on pairs of elements of a group, and we show that the rules for making the quilt of a Norton system from Sect. 1.3 hold rigorously, albeit with a few necessary corrections. We then provide some relatively small but useful examples of quilts in Chap. 5, concentrating on quilts of Norton systems.

We then proceed, in Chaps. 6 and 7, to our main results on the structure of quilts (not necessarily of Norton systems). Specifically, in Chap. 6, using elementary homology theory, we completely classify the subgroups of a Seifert group that project down to a given subgroup of the corresponding triangle group, and in Chap. 7, we interpret the results of Chap. 6 in terms of more standard versions of central extension theory. (For example, we recover the result of Conway, Coxeter, and Shephard [14] mentioned in Sect. 1.1.)

Part II of the monograph begins in Chap. 8 with our formulation of the *structure problem* in terms of combinatorial group theory. Roughly speaking, the structure problem asks: To what extent does a Norton system for a group G determine the structure of G? To give this problem a more rigorous framework, we define a concept called the *group of a quilt*, and we approach the structure problem by examining groups of quilts. Consequently, the majority of Chap. 8 is devoted to finding various algorithms for presenting the group of a quilt; in particular, the homology results of Chap. 6 are used to find particularly concise presentations. In Chap. 9, we then use the techniques of Chap. 8 to enumerate the groups of some small quilts.

Next, in Chap. 10, we consider the *monodromy action* of $\mathbf{B}_n$ on n-tuples of group elements, and we show that the monodromy action of $\mathbf{B}_3$ on triples of involutions of a group is essentially equivalent to Norton's action on pairs of elements. In Chap. 11, we then look at examples of the monodromy action on triples of *6-transpositions* (a class of involutions C such that for $x, y \in C$, the order of $xy \le 6$) in the Monster. Triples of 6-transpositions are notable because almost all of the corresponding quilts have genus zero. (Indeed, this "genus zero property" is one reason Norton first investigated quilts.)

In Chap. 12, we proceed to obtain assorted results on the structure problem. The highlights of this chapter include examples of finite quilts with infinite groups. Finally, in Chap. 13, we discuss some related recent work, and we also discuss some open questions and directions for further research.

In App. A, we discuss the problem of finding *independent generators* for a subgroup Γ of $\mathbf{PSL}_2(\mathbf{Z})$, that is, generators such that Γ is the free product of the corresponding cyclic groups. We give two solutions, one of which relies only on the *Reidemeister-Schreier method* from combinatorial group theory, and the other of which relies on quilts, especially the results of Chap. 8. For completeness, we also include a graphical description of Reidemeister-Schreier based on lectures given by Conway.

Logical structure. Logically, this monograph has the following structure, where $A \to B$ means that chapter B depends on chapter A:

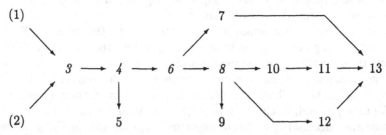

Chapters in italics (3, 4, 6, and 8) form the backbone of the monograph, chapters in parentheses (1 and 2) are introductory and contain only background material, and other chapters (5, 7, 9–13) contain other examples and results. As mentioned above, the appendix relies on parts of Chaps. 3 and 8, but may otherwise be read independently of the rest of the monograph.

2. Background material

In this chapter, we review some background material we will need later on (Sects. 2.2–2.7). We mostly keep matters informal, as our purpose is to review the basic facts for the reader who is familiar with this material, and to give the reader who is unfamiliar with this material enough intuition to follow the sequel. However, we do include proofs of non-standard results, and standard references are given as much as possible. We also set some notation (Sect. 2.1) and give a brief guide to the literature on background material that we do not otherwise attempt to describe (Sect. 2.8).

The reader who is familiar with most of this material may still wish to look at Sects. 2.2–2.4, since the somewhat non-standard notation and terminology introduced there is used frequently in the rest of the monograph.

2.1 Notation

In this section, we summarize some notation we will use throughout.

$\mathbf{Z}$ denotes the integers, $\mathbf{Q}$ denotes the rational numbers, $\mathbf{R}$ denotes the real numbers, and $\mathbf{C}$ denotes the complex numbers. $\mathbf{H}^2$ denotes the complex upper half-plane, and the element $\begin{pmatrix} a & b \\ c & d \end{pmatrix}$ of $\mathbf{PSL}_2(\mathbf{R})$ acts on $\mathbf{H}^2$ (on the left) by sending z to $(az + b)/(cz + d)$. $\mathbf{E}^2$ denotes the Euclidean plane, and $\mathbf{S}^n$ denotes the n-sphere.

We use the following group-theoretic terms and conventions. If $\alpha, \beta, \gamma, \ldots$ are elements of a group G, we use $\langle \alpha, \beta, \gamma, \ldots \rangle$ to denote the subgroup of G that they generate. If G acts on a set Ω, and $g \in G$, then by abuse of notation, we call an orbit of $\langle g \rangle$ on Ω a g-orbit. For groups H, N, and G, $H \leq G$ means that H is a subgroup of G, and $N \lhd G$ means that N is a normal subgroup of G. An *involution* is an element of order 2.

For group elements x, y, the *commutator* $[x, y] = x^{-1}y^{-1}xy$, and $x^y = y^{-1}xy$ denotes x conjugated on the right by y. The *derived series*, or *upper central series*, of a group G is the series $G^{(n)}$, where we define $G^{(0)}$ to be G and $G^{(n+1)}$ to be the commutator subgroup of $G^{(n)}$. We sometimes write $G^{(1)}$ as G', $G^{(2)}$ as G'', and so on. Similarly, the *lower central series* of G is the series $G_{(n)}$, where we define $G_{(0)}$ to be G and $G_{(n+1)}$ to be $[G, G_{(n)}]$. Note that our numbering of the lower central series differs by 1 from some other authors' notations; our only excuse is that neither notation seems to be standard.

We arbitrarily fix the convention that permutation group actions are *right* actions. For example, $(0\ 1) \cdot (1\ 2) = (0\ 2\ 1)$.

We use ATLAS [16] notation for group extensions. For example, $G \cong N \cdot H$ means that G is a group with a normal subgroup N such that $G/N \cong H$. Similarly, $G \cong N : H$ denotes a split extension (semidirect product), and $G \cong N \cdot H$ denotes a nonsplit extension.

The center of G is denoted by $Z(G)$, and the centralizer of $g \in G$ is denoted by $C_G(g)$. If G acts on a set Ω, then for $x \in \Omega$, $\text{Stab}_G(x)$ denotes the stabilizer of x in G. A *central extension* of a group H is defined to be a group $G \cong Z \cdot H$ such that $Z \le Z(G)$. $G * H$ denotes the free product of G and H. If H is a permutation group of degree n, $G \wr H$ denotes the semidirect product $G^n : H$, where the action of H permutes the coordinates of the Cartesian product G^n.

We use the following names for some particular finite groups. C_n denotes the cyclic group of order n, and the elements of C_n are generally written as integers mod n. We sometimes use the "topologist's notation" $\mathbf{Z}/M$ for the integers mod M. We also sometimes abbreviate C_n as n inside group shapes; for example, $2^4.5$ denotes a group G with a normal elementary abelian subgroup N of order 2^4 such that $G/N \cong C_5$. D_n denotes the dihedral group of order n. (In other words, the n in D_n is always even.) S_n denotes the group of permutations on n objects, and similarly, A_n denotes the group of even permutations on n objects. $W(E_n)$ denotes the Weyl group of the root system E_n. F_q denotes the field of q elements, and $L_n(q)$ (resp. $U_n(q)$) is the $n \times n$ projective special linear (resp. unitary) group over F_q. The symbol $\mathbf{M}$ denotes the Fischer-Griess Monster (the largest sporadic finite simple group). When in doubt, for finite groups, we follow the notation of the ATLAS [16], including the names of groups, outer automorphisms, and conjugacy classes.

We use revolutions, or *revs*, for short, as our units of angle and curvature, where 1 rev = 2π.

2.2 Triangle groups

In this section, we review some basic facts about triangle groups, along with a result (Thm. 2.2.5) we need for Sect. 7.2. Our standard references are Coxeter and Moser [22, §4.3, 4.4, 5.3] and Lyndon and Schupp [53, III.4–III.7].

Definition 2.2.1. For $m_1, m_2, n \ge 2$, the *triangle group* (m_1, m_2, n) is defined to be the group given by the presentation

$$\langle V_1, V_2, L \mid 1 = V_1^{m_1} = V_2^{m_2} = L^n = V_1 V_2 L \rangle. \tag{2.2.1}$$

Similarly, for $m_1, m_2 \ge 2$, the triangle group (m_1, m_2, ∞) is defined to be

$$\langle V_1, V_2, L \mid 1 = V_1^{m_1} = V_2^{m_2} = V_1 V_2 L \rangle. \tag{2.2.2}$$

Note that by conjugating both sides of $1 = V_1 V_2 L$ by V_1, we see that the group (m_1, m_2, n) is isomorphic to the group (m_2, n, m_1). Similarly, by inverting both sides of $1 = V_1 V_2 L$, we see that (m_1, m_2, n) is isomorphic to (n, m_1, m_2). It follows that the isomorphism class of (m_1, m_2, n) is independent of the order in which m_1, m_2, and n are written. Nevertheless, it will later be convenient for us to treat the last generator L and the last exponent n separately.

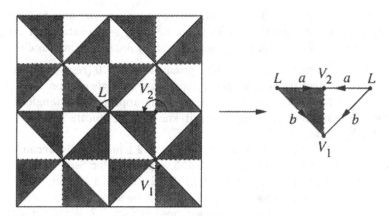

Fig. 2.2.1. The triangle group $(4, 2, 4)$

The group (m_1, m_2, n) is called a triangle group because it is best understood by its faithful action as a discrete group of isometries of a geometric space, with quotient a "triangular pillowcase." For instance, consider Fig. 2.2.1. If we let V_1, V_2, and L be the Euclidean isometries indicated on the left-hand side of Fig. 2.2.1 (counterclockwise turns of $\frac{1}{4}$ rev, $\frac{1}{2}$ rev, and $\frac{1}{4}$ rev, respectively), then $\langle V_1, V_2, L \mid 1 = V_1^4 = V_2^2 = L^4 = V_1 V_2 L \rangle$, the triangle group $(4, 2, 4)$, surjects onto the group $\Gamma = \langle V_1, V_2, L \rangle$ of Euclidean isometries. In fact, it can be shown that this surjection is an isomorphism, that Γ is a discrete subgroup of $\mathrm{Isom}(\mathbf{E}^2)$, and that $\Gamma \backslash \mathbf{E}^2$ is the space O obtained by taking the triangle on the right-hand side of Fig. 2.2.1 and identifying the edges marked a and b as indicated. As mentioned above, the resulting quotient looks like a sealed flat triangular pillowcase.

Now, even though topologically, O is just $\mathbf{S}^2$, geometrically, the corners of O are different from the other points of O. More precisely, a small circle around the corner of O marked V_2 only measures out an angle of $\frac{1}{2}$ rev, which corresponds to the fact that the stabilizer of a preimage of that corner is an order 2 subgroup of Γ conjugate to $\langle V_2 \rangle$. We call the corner V_2 a *cone point of order 2*. Similarly, a small circle around one of the other corners of O only measures out an angle of $\frac{1}{4}$ rev, which corresponds to the fact that the other corners come from points fixed by subgroups conjugate to either $\langle V_1 \rangle$ or $\langle L \rangle$. In other words, the other corners of O are cone points of order 4. This space

O, along with its cone point structure, is an example of an *orbifold*, that is, a manifold with singularities such as cone points.

It turns out that an analogous construction works for all triangle groups. In fact, there is nothing special about $(4, 2, 4)$, except that it acts naturally on the Euclidean plane $\mathbf{E}^2$, and not the sphere $\mathbf{S}^2$ or the hyperbolic plane $\mathbf{H}^2$. In general, we have:

Theorem 2.2.2. *For $m_1, m_2, n \geq 2$, let $\chi = \dfrac{1}{m_1} + \dfrac{1}{m_2} + \dfrac{1}{n} - 1$. Then for $\chi > 0$ (resp. $\chi = 0$, $\chi < 0$), the triangle group (m_1, m_2, n) acts faithfully as a discrete group of isometries of $\mathbf{S}^2$ (resp. $\mathbf{E}^2$, $\mathbf{H}^2$) with quotient an orbifold that is topologically $\mathbf{S}^2$ and has 3 cone points, of order m_1, m_2, and n.* ☐

Because of Thm. 2.2.2, the triangle groups with $\chi > 0$ (resp. $\chi = 0$, $\chi < 0$) are called *spherical* (resp. *Euclidean, hyperbolic*) triangle groups.

Thm. 2.2.2 may be proved by constructing appropriate triangles in $\mathbf{S}^2$, $\mathbf{E}^2$, or $\mathbf{H}^2$, and then applying a result known as *Poincaré's theorem*. See Maskit [55] for details.

We have introduced Thm. 2.2.2 and Fig. 2.2.1 here partly because they provide geometric motivation for Chap. 3, especially for Sects. 3.2 and 3.5. (Compare, for instance, Figs. 2.2.1, 3.2.1, 3.2.2, and 3.5.1.) In addition, using Thm. 2.2.2 and a well-known fact (Prop. 2.2.4), we also obtain a result about triangle groups (Thm. 2.2.5) that we will need in Sect. 7.2. We begin with the following theorem, which is not obvious from presentation (2.2.1), but follows clearly from the geometric picture. (More rigorously, this is essentially a result of Poincaré's theorem; see Maskit [55].)

Proposition 2.2.3. *Let Γ be the group defined by presentation (2.2.1). Then any element $\gamma \in \Gamma$ of finite order is conjugate to an element of $\langle V_1 \rangle$, $\langle V_2 \rangle$, or $\langle L \rangle$.*

Idea of proof. Let X be the space on which Γ acts in the statement of Thm. 2.2.2 (that is, $X = \mathbf{S}^2$, $\mathbf{E}^2$, or $\mathbf{H}^2$). Since any element of $\text{Isom}(X)$ of finite order must fix a point of X, γ must fix a point $x \in X$, which means that the image of x in $O = \Gamma \backslash X$ must be a cone point of O. However, the stabilizer of a point in the inverse image of a cone point of O must be conjugate to a subgroup of $\langle V_1 \rangle$, $\langle V_2 \rangle$, or $\langle L \rangle$. ☐

We also recall:

Proposition 2.2.4. *For $m_1, m_2, n \geq 2$, there exists a finite group G and elements $A, B, C \in G$ such that the orders of A, B, and C are precisely m_1, m_2, and n, respectively, and $1 = ABC$.* ☐

For instance, we may take G to be $L_2(q)$ for some sufficiently large prime power q (Feuer [32]). In any case, we consequently obtain the following fact, to be used in Sect. 7.2. (Note that the last part of the proof of Thm. 2.2.5 relies on some topology that we will not discuss explicitly; for details, see, for instance, Massey [57].)

Theorem 2.2.5. *The triangle group* (m_1, m_2, n) $(m_1, m_2, n \geq 2)$ *contains a torsion-free subgroup Γ of finite index. Furthermore, if (m_1, m_2, n) is Euclidean or hyperbolic, then the abelianization of Γ is a nontrivial free abelian group.*

Proof. Let A, B, and C be the elements of the finite group G from Prop. 2.2.4, and let Γ be the kernel of the homorphism from (m_1, m_2, n) to G that sends V_1 to A, V_2 to B, and L to C. If γ were a nontrivial element of Γ of finite order, then the element of (say) $\langle V_1 \rangle$ conjugate to γ would also be contained in Γ, which would imply that the order of A is $< m_1$; contradiction. The first statement of the theorem follows.

As for the second statement, suppose that (m_1, m_2, n) is Euclidean or hyperbolic, and let X be the space on which (m_1, m_2, n) naturally acts, as in the statement of Thm. 2.2.2. Since Γ is torsion-free of finite index in the infinite group (m_1, m_2, n), and non-torsion elements of a discrete subgroup of $\mathrm{Isom}(X)$ are fixed-point free, we know from the theory of covering spaces (see Hatcher [39] or Massey [57, Chap. V]) that $\Gamma \backslash X$ is a connected compact surface with infinite fundamental group Γ. By the classification of compact surfaces (Massey [57, Chap. I]), the abelianization of Γ is nontrivial free abelian. $\qquad\square$

2.3 Seifert groups

The groups we consider in this section have been investigated in Conway, Coxeter, and Shephard [14], Coxeter and Moser [22, §6.5–6.7], and in a topological context, in Seifert [79] (see also Sect. 7.3). Our notation is essentially the notation from Conway, Coxeter, and Shephard [14], rearranged slightly, so we take that as our standard reference.

We begin with two definitions.

Definition 2.3.1. For $m_1, m_2, n \geq 2$, and for $p_1, p_2, q, s \in \mathbf{Z}$, we define the *Seifert group* $\left\langle \frac{p_1}{m_1}, \frac{p_2}{m_2}, \frac{q}{n} \right\rangle_s$ to be the group given by the presentation

$$\left\langle Z, V_1, V_2, L \,\middle|\, \begin{array}{l} 1 = [Z, V_1] = [Z, V_2] = [Z, L], \\ Z^{p_1} = V_1^{m_1}, \; Z^{p_2} = V_2^{m_2}, \; Z^q = L^n, \; Z^s = V_1 V_2 L \end{array} \right\rangle. \quad (2.3.1)$$

Continuing analogously to Defn. 2.2.1, in the case "$n = \infty$", we see that the value of s does not affect the isomorphism type of the group (replace L with LZ^s). We therefore define the Seifert group $\left\langle \frac{p_1}{m_1}, \frac{p_2}{m_2}, 0 \right\rangle$ to be

$$\left\langle Z, V_1, V_2, L \,\middle|\, \begin{array}{l} 1 = [Z, V_1] = [Z, V_2] = [Z, L], \\ Z^p = V_1^{m_1}, \; Z^q = V_2^{m_2}, \; 1 = V_1 V_2 L \end{array} \right\rangle. \quad (2.3.2)$$

By convention, in any Seifert group, we also define $R = V_2 V_1$.

Note that we can *not* reduce the "fractions" in $\left\langle \frac{p_1}{m_1}, \frac{p_2}{m_2}, \frac{q}{n} \right\rangle_s$ to lowest terms and still have the same group. On the other hand, the minus signs in something like $\left\langle -\frac{p_1}{m_1}, -\frac{p_2}{m_2}, -\frac{q}{n} \right\rangle_{-s}$ are unambiguous, since it does not matter whether the signs are in the "numerator" or "denominator" of a "fraction." By convention, we take the "denominator" to be positive.

Definition 2.3.2. We define the *Euler number* $e(\Sigma) \in \mathbf{Q}$ of the Seifert group $\Sigma = \left\langle \frac{p_1}{m_1}, \frac{p_2}{m_2}, \frac{q}{n} \right\rangle_s$ to be

$$e(\Sigma) = \frac{p_1}{m_1} + \frac{p_2}{m_2} + \frac{q}{n} - s, \qquad (2.3.3)$$

and similarly, we define the Euler number of $\Sigma = \left\langle \frac{p_1}{m_1}, \frac{p_2}{m_2}, 0 \right\rangle$ to be

$$e(\Sigma) = \frac{p_1}{m_1} + \frac{p_2}{m_2}. \qquad (2.3.4)$$

For motivation behind the name "Euler number", see Scott [78, §1–3].

Let $\Sigma = \left\langle \frac{p_1}{m_1}, \frac{p_2}{m_2}, \frac{q}{n} \right\rangle_s$. As with triangle groups, by conjugating the relation $Z^s = V_1 V_2 L$ by V_1, we see that

$$\Sigma \cong \left\langle \frac{p_2}{m_2}, \frac{q}{n}, \frac{p_1}{m_1} \right\rangle_s, \qquad (2.3.5)$$

and by inverting all non-commutator relations and all generators, we see that

$$\Sigma \cong \left\langle \frac{q}{n}, \frac{p_2}{m_2}, \frac{p_1}{m_1} \right\rangle_s. \qquad (2.3.6)$$

In other words, as with triangle groups, the isomorphism class of Σ is independent of the order in which $\frac{p_1}{m_1}$, $\frac{p_2}{m_2}$, and $\frac{q}{n}$ are written. Furthermore, if we define $-\Sigma$ to be the group obtained by replacing the generator Z with its inverse, then $-\Sigma = \left\langle -\frac{p_1}{m_1}, -\frac{p_2}{m_2}, -\frac{q}{n} \right\rangle_{-s}$, $e(-\Sigma) = -e(\Sigma)$, and $-\Sigma \cong \Sigma$.

Note that by replacing V_1 (say) with $V_1 Z^{-1}$, we obtain

$$\left\langle \frac{p_1}{m_1}, \frac{p_2}{m_2}, \frac{q}{n} \right\rangle_s \cong \left\langle \frac{p+m_1}{m_1}, \frac{p_2}{m_2}, \frac{q}{n} \right\rangle_{s+1}. \qquad (2.3.7)$$

More generally, up to isomorphism, we may add any integers to $\frac{p_1}{m_1}$, $\frac{p_2}{m_2}$, $\frac{q}{n}$, and s, as long as the Euler number of the group remains constant. Consequently, for most purposes, we may assume $s = 0$, and we write $\left\langle \frac{p_1}{m_1}, \frac{p_2}{m_2}, \frac{q}{n} \right\rangle_0$ as $\left\langle \frac{p_1}{m_1}, \frac{p_2}{m_2}, \frac{q}{n} \right\rangle$.

We are particularly interested in the following class of Seifert groups.

Definition 2.3.3. For $n = \infty$, a Seifert group $\left\langle \frac{p_1}{m_1}, \frac{p_2}{m_2}, 0 \right\rangle$ is said to be *geometric* if $\gcd(p_r, m_r) = 1$ for $r = 1, 2$. For $n < \infty$, a Seifert group $\left\langle \frac{p_1}{m_1}, \frac{p_2}{m_2}, \frac{q}{n} \right\rangle$ is said to be *geometric* if $\gcd(p_r, m_r) = \gcd(q, n) = 1$ for $r = 1, 2$.

Remark 2.3.4. The Seifert groups in Defn. 2.3.3 are called geometric because they are precisely the Seifert groups that arise from Seifert fibered 3-manifolds. See Sect. 7.3 for more on this topic.

Seifert groups are closely related to triangle groups. Specifically, by killing Z, we see that the Seifert group $\Sigma = \left\langle \frac{p_1}{m_1}, \frac{p_2}{m_2}, \frac{q}{n} \right\rangle$ is a central extension of the triangle group (m_1, m_2, n) by the cyclic subgroup $\langle Z \rangle$ of Σ.

As we will show (Thm. 6.1.6), the Euler number of a Seifert group $\Sigma = \left\langle \frac{p_1}{m_1}, \frac{p_2}{m_2}, \frac{q}{n} \right\rangle$ describes how Σ differs from the direct product $(m_1, m_2, n) \times \langle Z \rangle$. For now, we illustrate the importance of $e(\Sigma)$ by proving Thm. 2.3.6 (following Seifert [79, §12]) for use in Sect. 7.2.

First, to state Thm. 2.3.6, we need the following definition.

Definition 2.3.5. A group G is *perfect* if $G' = G$, or in other words, if the abelianization of G is trivial.

Note that any quotient of a perfect group is perfect (exercise).

Theorem 2.3.6 (Seifert). *For $m_1, m_2, n \geq 2$, we have:*

1. *The triangle group (m_1, m_2, n) is perfect if and only if m_1, m_2, and n are pairwise relatively prime.*

2. *For $p_1, p_2, q, s \in \mathbf{Z}$, the Seifert group $\Sigma = \left\langle \frac{p_1}{m_1}, \frac{p_2}{m_2}, \frac{q}{n} \right\rangle_s$ is perfect if and only if m_1, m_2, and n are pairwise relatively prime and $|e(\Sigma)| = \frac{1}{m_1 m_2 n}$.*

3. *For m_1, m_2, and n pairwise relatively prime, there exists a perfect Seifert group $\left\langle \frac{p_1}{m_1}, \frac{p_2}{m_2}, \frac{q}{n} \right\rangle_s$, and this group is unique up to isomorphism.*

Proof. First, if (for instance) $\gcd(m_1, m_2) = d > 1$, then by setting $L = 1$ and $V_1 = V_2^{-1}$ in $\langle V_1, V_2, L \,|\, 1 = V_1^{m_1} = V_2^{m_2} = L^n = V_1 V_2 L \rangle$, we obtain a cyclic quotient of (m_1, m_2, n) of order d. Therefore, since any quotient of a perfect group is perfect, it remains only to prove assertions (2) and (3) in the case where m_1, m_2, and n are pairwise relatively prime.

So let $\Sigma = \left\langle \frac{p_1}{m_1}, \frac{p_2}{m_2}, \frac{q}{n} \right\rangle_s$ for m_1, m_2, and n pairwise relatively prime. Abelianizing presentation (2.3.1) and writing group elements additively, we get

$$
\begin{aligned}
m_1 V_1 \qquad\qquad - p_1 Z &= 0 \\
m_2 V_2 \qquad - p_2 Z &= 0 \\
n L - q Z &= 0 \\
V_1 + \quad V_2 + \quad L - \quad s Z &= 0.
\end{aligned}
\tag{2.3.8}
$$

Since we have 4 generators and 4 relations, this quotient is trivial if and only if the determinant of the matrix

$$
\begin{pmatrix}
m_1 & & & -p_1 \\
& m_2 & & -p_2 \\
& & n & -q \\
1 & 1 & 1 & -s
\end{pmatrix}
\tag{2.3.9}
$$

is equal to ± 1. A calculation then shows that the determinant of (2.3.9) is

$$
m_1 m_2 n \cdot \left(\frac{p_1}{m_1} + \frac{p_2}{m_2} + \frac{q}{n} - s \right) = m_1 m_2 n \cdot e(\Sigma).
\tag{2.3.10}
$$

Assertion (2) follows.

As for assertion (3), recall from the discussion after Defn. 2.3.2 that, up to isomorphism, we may assume that $e(\Sigma) > 0$ and $0 < \frac{p_1}{m_1}, \frac{p_2}{m_2}, \frac{q}{n} < 1$. Under these assumptions, assertion (2) implies that $\left\langle \frac{p_1}{m_1}, \frac{p_2}{m_2}, \frac{q}{n} \right\rangle_s$ is perfect if and only if

$$
p_1 m_2 n + p_2 m_1 n + q m_1 m_2 = s m_1 m_2 n + 1.
\tag{2.3.11}
$$

Now, since m_1, m_2, and n are pairwise relatively prime, by the Chinese Remainder Theorem, we may solve (2.3.11) for p_1, p_2, and q, given any value of s. On the other hand, if p_1, p_2, q, s and p_1', p_2', q', s' are both solutions to (2.3.11), then

$$
(p_1 - p_1') m_2 n + (p_2 - p_2') m_1 n + (q - q') m_1 m_2 = (s - s') m_1 m_2 n,
\tag{2.3.12}
$$

which means that $p_1 \equiv p_1' \pmod{m_1}$, $p_2 \equiv p_2' \pmod{m_2}$, and $q \equiv q' \pmod{n}$. Since we are assuming $0 < \frac{p_1}{m_1}, \frac{p_2}{m_2}, \frac{q}{n} < 1$, the solutions must actually be identical. The theorem follows. □

2.4 Braid groups

In this section, we present a brief review of some basic facts about braid groups. Our standard reference is Birman [6].

The *n-string braid group*, denoted by $\mathbf{B}_n$, is the group of all *isotopy classes* of *braid moves* on n strings. By braid move, we mean a string-tying pattern like the ones in Figs. 2.4.1–2.4.3, and by isotopy classes, we mean the equivalence classes of braid moves under the relation of continously deforming one move into another while leaving the ends of the strings fixed. The product of two braid moves is defined to be the move consisting of one braid move, and then the other, with proceeding down the strings corresponding to composing on the right, by convention. (This informal description may be formalized by

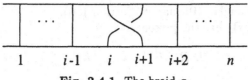

Fig. 2.4.1. The braid σ_i

defining $\mathbf{B}_n$ as the fundamental group of a certain configuration space of n points in the plane; see Birman [6, Sect. 1.1].)

For instance, let σ_i be the braid move shown in Fig. 2.4.1, that is, braiding string i over string $i + 1$. Then the left-hand braid in Fig. 2.4.2 is $\sigma_1\sigma_2\sigma_1$, and the right-hand braid is $\sigma_2\sigma_1\sigma_2$. We suggest that the reader who is not familiar with braid groups check, by manipulating some pieces of string, that these braids are isotopic, yielding the relation $\sigma_1\sigma_2\sigma_1 = \sigma_2\sigma_1\sigma_2$. Similarly, the left-hand braid in Fig. 2.4.3 is $\sigma_1\sigma_3$, the right-hand braid is $\sigma_3\sigma_1$, and again, the braids are isotopic, yielding the relation $\sigma_1\sigma_3 = \sigma_3\sigma_1$.

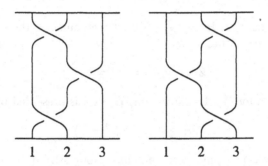

Fig. 2.4.2. A pair of isotopic braids

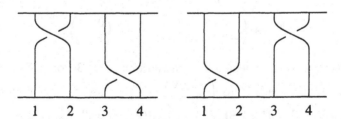

Fig. 2.4.3. Another pair of isotopic braids

In fact, the σ_i generate $\mathbf{B}_n$, and

$$\left\langle \sigma_1, \ldots, \sigma_{n-1} \;\middle|\; \begin{array}{l} \sigma_i\sigma_{i+1}\sigma_i = \sigma_{i+1}\sigma_i\sigma_{i+1}, \\ [\sigma_i, \sigma_j] = 1 \text{ for } |i - j| \geq 2 \end{array} \right\rangle \qquad (2.4.1)$$

is a presentation for $\mathbf{B}_n$ (Birman [6, Section 1.4]).

In particular, $\mathbf{B}_3$ has the presentation

$$\langle L, R \,|\, LR^{-1}L = R^{-1}LR^{-1} \rangle, \tag{2.4.2}$$

where $L = \sigma_1$ and $R = \sigma_2^{-1}$. Note that it follows that there is an (outer) automorphism of $\mathbf{B}_3$ that exchanges L and R (invert both sides of the relation $LR^{-1}L = R^{-1}LR^{-1}$).

If we now let $V_1 = R^{-1}L$, $V_2 = RL^{-1}R$, and $Z = V_1^3$, (2.4.2) implies that

$$Z = R^{-1}LR^{-1}LR^{-1}L = (R^{-1}LR^{-1})^2 = V_2^{-2}. \tag{2.4.3}$$

Since $R = V_2V_1$ is then redundant, and the defining relation of presentation (2.4.2) is equivalent to $V_1V_2L = 1$, we see that

$$\mathbf{B}_3 \cong \left\langle Z, V_1, V_2, L \,\middle|\, \begin{matrix} 1 = [Z, V_1] = [Z, V_2] = [Z, L], \\ Z = V_1^3, \ Z^{-1} = V_2^2, \ 1 = V_1V_2L \end{matrix} \right\rangle, \tag{2.4.4}$$

or in other words, that $\mathbf{B}_3 \cong \langle \frac{1}{3}, -\frac{1}{2}, 0 \rangle$. By eliminating the redundant generator L, we also see that

$$\langle Z, V_1, V_2 \,|\, Z = V_1^3 = V_2^{-2} \rangle \tag{2.4.5}$$

is a presentation for $\mathbf{B}_3$. It follows that $\mathbf{B}_3/\langle Z \rangle$ is presented by

$$\langle V_1, V_2 \,|\, V_1^3 = V_2^{-2} = 1 \rangle, \tag{2.4.6}$$

and is therefore isomorphic to the modular group $\mathbf{PSL}_2(\mathbf{Z})$. In fact, we may choose the projection from $\mathbf{B}_3$ to $\mathbf{PSL}_2(\mathbf{Z})$ such that

$$L \mapsto \begin{pmatrix} 1 & 1 \\ 0 & 1 \end{pmatrix}, \qquad R \mapsto \begin{pmatrix} 1 & 0 \\ 1 & 1 \end{pmatrix},$$

$$V_1 \mapsto \begin{pmatrix} 1 & 1 \\ -1 & 0 \end{pmatrix}, \qquad V_2 \mapsto \begin{pmatrix} 0 & -1 \\ 1 & 0 \end{pmatrix}. \tag{2.4.7}$$

More precisely, (2.4.7) defines a homomorphism of $\mathbf{B}_3$ onto $\mathbf{SL}_2(\mathbf{Z})$ with kernel $\langle Z^2 \rangle$ that sends Z to $\begin{pmatrix} -1 & 0 \\ 0 & -1 \end{pmatrix}$. In any case, by convention, when working with either $\mathbf{B}_3$, $\mathbf{PSL}_2(\mathbf{Z})$, or $\mathbf{SL}_2(\mathbf{Z})$, we will always use L, R, V_1, V_2, and Z in the manner indicated above.

Note that since much of this monograph deals with Seifert groups in general, and not just $\mathbf{B}_3$, we use presentation (2.4.4) as our standard presentation of $\mathbf{B}_3$, in contrast with previous versions of some of this material.

2.5 Some elementary homology theory

In this section, we review some basic definitions and results in algebraic topology, concentrating on the homology and cohomology of closed orientable surfaces. The material here is somewhat involved, and is used only in a few crucial places (Chaps. 3, 6, and 8), so the reader may prefer to skip this section initially and refer back as necessary. Our standard references are Hatcher [39] and Munkres [60].

We begin by defining the class of topological spaces we will consider.

Definition 2.5.1. We define a *2-complex* to be a topological space X constructed by the following procedure. Let I denote the unit interval, and for $m \geq 3$, let P_m denote a fixed regular m-gon of side length 1 in $\mathbf{R}^2$, along with its interior.

1. We start with a discrete set X^0, also known as the *0-cells*, the *0-skeleton*, or the *vertices* of X.

2. We form the *1-skeleton* X^1 of X by gluing a collection $\{I_\alpha\}$ of copies of I onto X^0 using attaching maps $\varphi_\alpha : \partial I_\alpha \to X^0$, where ∂I_α denotes the boundary (the endpoints) of the interval I_α. In other words, we take the disjoint union of X^0 and $\{I_\alpha\}$, choose maps $\varphi_\alpha : \partial I_\alpha \to X^0$, and form the quotient space X^1, with the quotient topology, under the equivalence relation $x \sim \varphi_\alpha(x)$. The images of the I_α in the quotient are called the (closed) *1-cells* or the *edges* of X. Note that the interior of each 1-cell of X inherits a metric structure from I.

3. For each β in some index set, let P_β be a copy of some polygon P_m. We form X by gluing the disjoint union $\{P_\beta\}$ onto X^1 using attaching maps $\rho_\beta : \partial P_\beta \to X^1$ such that

   a) ρ_β sends each vertex of P_β to a 0-cell of X^1; and

   b) the restriction of ρ_β to the interior of each of the sides of ∂P_β is an isometry from the interior of that side to the interior of a 1-cell of X^1.

   The images of the P_β in the quotient are called the *2-cells* of X.

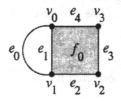

Fig. 2.5.1. The 2-complex X_1

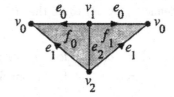

Fig. 2.5.2. The 2-complex X_2

For example, graphs (possibly with multiple edges or self-loops) are precisely those 2-complexes with no 2-cells. As for 2-dimensional examples, consider the 2-complexes X_1 and X_2 shown in Figs. 2.5.1 and 2.5.2, respectively. X_1 is constructed by taking the 0-cells v_0, v_1, v_2, and v_3; gluing both e_0 and e_1 to v_0 and v_1, e_2 to v_1 and v_2, e_3 to v_2 and v_3, and e_4 to v_3 and v_0; and gluing the square f_0 to e_1, e_2, e_3, and e_4. Similarly, if X_2 is the 2-complex obtained by taking Fig. 2.5.2 and identifying the two copies of e_0, as shown, and the two copies of e_1, as shown, then X_2 is constructed by taking the 0-cells v_0, v_1, and v_2; gluing e_0 to v_0 and v_1, e_1 to v_0 and v_2, and e_2 to v_1 and v_2; and gluing the triangles f_0 and f_1 to e_0, e_1, and e_2. Note that X_2 is homeomorphic to $\mathbf{S}^2$.

Next, we define the invariants known as *homology* and *cohomology*. In Defns. 2.5.2–2.5.7, let X be a 2-complex.

| Fig. 2.5.3. Oriented 1-cell | Fig. 2.5.4. Oriented 2-cells |

Definition 2.5.2. An *oriented 0-cell* of X is defined to be a 0-cell of X together with a sign $+$ or $-$.

An *oriented 1-cell* of X is defined to be a 1-cell e of X together with a choice of "direction of travel" along e (Fig. 2.5.3). More precisely, in the notation of Defn. 2.5.1, if I_α is the interval glued onto X^0 to form e, an orientation on e is a choice of direction (left or right) on I_α.

An *oriented 2-cell* of X is defined to be a 2-cell f of X together with a choice of "cyclic direction of travel" around f (Fig. 2.5.4). More precisely, in the notation of Defn. 2.5.1, if P_β is the polygon glued onto X^1 to form f, an orientation on f is a choice of either clockwise or counterclockwise on P_β.

If σ is an oriented n-cell, then $-\sigma$ denotes the same n-cell with the opposite orientation.

Definition 2.5.3. Let R be a commutative ring with identity. For $n = 0, 1, 2$, we define $C_n(X; R)$ to be the R-module of functions c from the oriented n-cells of X to R such that

$$c(-\sigma) = -c(\sigma) \quad \text{for any oriented } n\text{-cell } \sigma, \text{ and} \tag{2.5.1}$$
$$c(\sigma) = 0 \quad \text{for all but finitely many oriented } n\text{-cells } \sigma, \tag{2.5.2}$$

with the sum of $c_1, c_2 \in C_n(X; R)$ defined to be the function $c_1(\sigma) + c_2(\sigma)$. In other words, if we identify any oriented n-cell σ of X with the element of $C_n(X; R)$ that takes the value 1 on σ, -1 on $-\sigma$, and 0 on every other oriented n-cell of X, then $C_n(X; R)$ is the free R-module with double basis the oriented n-cells of X. The elements of $C_n(X; R)$ are called the *n-chains of X with values in R.* For $n < 0$ or $n > 2$, we define $C_n(X; R)$ to be the 0 module.

Definition 2.5.4. For $n = 0, 1, 2$, we define a module homomorphism ∂_n from $C_n(X; R)$ to $C_{n-1}(X; R)$ by setting

$$\partial_0 v = 0 \tag{2.5.3}$$

for every oriented 0-cell v;

$$\partial_1 e = v_1 - v_0 \tag{2.5.4}$$

for every oriented 1-cell e that is glued to v_0 and v_1, and travels from v_0 to v_1, as shown in Fig. 2.5.5; and

$$\partial_2 f = e_0 + \cdots + e_{m-1} \tag{2.5.5}$$

for every oriented 2-cell f that is glued to $e_0, \dots, e_{m-1}$, and travels around them as shown in Fig. 2.5.6. For $n < 0$ or $n > 2$, we define $\partial_n = 0$. The subscript on ∂_n will usually be clear from the context in which it is used, so it will generally be omitted.

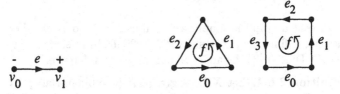

Fig. 2.5.5. Boundary map ∂_1 **Fig. 2.5.6.** Boundary maps ∂_2

Definition 2.5.5. The *n-cycles* of X with values in R, denoted by $Z_n(X; R)$, are the submodule of all $c \in C_n(X; R)$ such that $\partial c = 0$. The *n-boundaries* of X with values in R, denoted by $B_n(X; R)$, are the submodule of all $c \in C_n(X; R)$ such that $c = \partial a$ for some $a \in C_{n+1}(X; R)$. At this point, we suggest that the reader who is less familiar with homology verify, using (2.5.3)–(2.5.5) and Figs. 2.5.5 and 2.5.6, that $\partial^2 = 0$, or in other words, that $B_n(X; R) \subset Z_n(X; R)$ for all n. (Geometrically, this just says that every n-boundary is an n-cycle, with $n = 1$ being the only non-trivial verification in our case.)

We may therefore define $H_n(X; R)$, the *nth homology group of X with values in R*, to be $Z_n(X; R)/B_n(X; R)$. Equivalently, we say that the sequence of homomorphisms

$$1 \longrightarrow B_n(X; R) \longrightarrow Z_n(X; R) \longrightarrow H_n(X; R) \longrightarrow 1$$

is *exact*, or in other words, that the image of each homomorphism in the sequence is precisely the kernel of the next.

Notation 2.5.6. By convention, when $R = \mathbf{Z}$, it is omitted. That is, $C_n(X) = C_n(X; \mathbf{Z})$, $H_n(X) = H_n(X; \mathbf{Z})$, and so on.

By "dualizing" Defns. 2.5.3–2.5.5, we obtain the *cohomology* of X. That is:

Definition 2.5.7. Let $C^n(X; R)$ denote the R-module of homomorphisms from $C_n(X)$ to R (the *n-cochains of X with values in R*). We define the map $\delta_n : C^n(X; R) \to C^{n+1}(X; R)$ by $\delta_n c(d) = c(\partial_{n+1} d)$, $B^n(X; R) = \delta_{n-1} C^{n-1}(X; R)$ (the *n-coboundaries of X*), and $Z^n(X; R) = \ker \delta_n$ (the *n-cocycles of X*). As before, the subscript on δ_n will generally be omitted. As in Defn. 2.5.5, $\delta^2 = 0$, so we may define $H^n(X; R)$, the *nth cohomology group of X with values in R*, to be $Z^n(X; R)/B^n(X; R)$.

We have the following general description of 0-homology and cohomology.

Theorem 2.5.8. *If X has c path-components, then $H_0(X; R) \cong H^0(X; R) \cong R^c$. In particular, if X is path-connected ($c = 1$), then a 0-chain of X is a 0-boundary if and only if the sum of its coefficients is 0.* □

2.6 Graphs and surfaces

Recall that a graph is simply a 2-complex with no 2-cells. We will need the following definitions and results (2.6.1–2.6.6) from elementary graph theory. We take Diestel [24, Chap. I] as our standard reference.

Definition 2.6.1. Let X be a graph. A (combinatorial) *path* in X is a sequence $(v_0, e_1, v_1, \ldots, e_n, v_n)$ such that e_i is a directed edge from v_{i-1} to v_i. A path in X is said to be *reducible* if for some i, e_{i+1} is e_i, traversed in the opposite direction; otherwise, the path is said to be *irreducible*. A *cycle* in X is a path that begins and ends at the same vertex. (Note that the sum of the directed edges in a cycle is a 1-cycle in X, in the sense of Defn. 2.5.5.) If there are no irreducible cycles in X, X is said to be a *tree*.

Theorem 2.6.2. *A finite graph with c components, e edges, and v vertices is a tree if and only if $v = e + c$. In particular, a connected graph is a tree if and only if $v = e + 1$.* □

Definition 2.6.3. We call a vertex of X *terminal* if precisely one edge is attached to it, and we call an edge of X *terminal* if it is attached to a terminal vertex. We call the operation of taking a graph and removing one of its terminal vertices and its attached terminal edge *pruning a terminal edge*.

Theorem 2.6.4. *Let X be a finite tree. If X has at least 3 vertices, then it has at least 2 distinct terminal edges. Furthermore, X may be reduced to a point by repeatedly pruning terminal edges.* □

Definition 2.6.5. We say that a subgraph T of a graph X is a *spanning tree* for X if T is a tree containing every vertex of X.

Theorem 2.6.6. *Every finite graph contains a spanning tree.*   □

We next consider surfaces, taking Massey [57, Chap. I] and Munkres [60] as our standard references. We begin by defining the particular surfaces in which we are interested.

Definition 2.6.7. A 2-complex X that is also a surface is called a *combinatorial surface*. If all of the 2-cells of X are triangles, then X is also called a *triangulated surface*. For example, the complex X_2 in Fig. 2.5.2, p. 19, is a triangulated sphere.

A *closed* surface is defined to be a compact connected surface without boundary. A closed combinatorial surface X is said to be *orientable* if the 2-cells of X may be oriented in a manner such that $\sum \sigma \in Z_2(X; \mathbf{Z})$, where the sum is taken over all 2-cells of X.

Fig. 2.6.1. Local picture of a consistent orientation

Let X be a closed combinatorial surface. Since every 2-cell of X must meet precisely one other 2-cell of X (possibly itself) at each of its edges, and every 1-cell of X must be a side of precisely two 2-cells (counting multiplicities), we see that X is orientable if and only if its 2-cells may be consistently oriented so every 1-cell meets its two faces like the 1-cell in the middle of Fig. 2.6.1 meets its two faces.

Recall that the *classification of closed orientable surfaces* (see Massey [57, Chap. 1]) states that any closed orientable surface is homeomorphic to the connected sum of g tori, where the connected sum of 0 tori is the sphere. Such a surface is called a *surface of genus g*. For example, Fig. 2.6.2 shows, from left to right, the connected sum of 0, 1, and 2 tori, or in other words, surfaces of genus 0, 1, and 2.

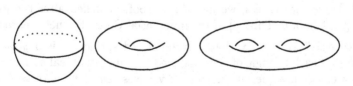

Fig. 2.6.2. Surfaces of genus 0, 1, and 2

Since homology is a topological invariant of a 2-complex (see, for instance, Munkres [60, §18]), the following theorem then implies that closed orientable surfaces are completely determined topologically by their genus.

Theorem 2.6.8. *Let X be a closed orientable surface of genus g. Then for any commutative ring R, $H_0(X;R) \cong H^0(X;R) \cong R$, $H_1(X;R) \cong H^1(X;R) \cong R^{2g}$, $H_2(X;R) \cong H^2(X;R) \cong R$, and $H_n(X;R) \cong H^n(X;R) \cong 0$ for $n > 2$.* □

We will also need the classical result known as *Euler's formula* (see Munkres [60, Thm. 22.2]). First, however, we need a definition.

Definition 2.6.9. Let X be a finite 2-complex, and let $\beta_n(X)$ (the *nth Betti number of X*) denote the rank (over **Z**) of $H_n(X)$ mod its torsion subgroup. The *Euler characteristic* $\chi(X)$ of X is defined to be

$$\chi(X) = \beta_0(X) - \beta_1(X) + \beta_2(X). \tag{2.6.1}$$

Theorem 2.6.10 (Euler's formula). *If X is a 2-complex, then*

$$\chi(X) = V - E + F, \tag{2.6.2}$$

where V, E, and F are the number of 0-cells, 1-cells, and 2-cells of X, respectively. □

In particular, since Thm. 2.6.8 implies that $\chi(X) = 2 - 2g$ for a surface of genus g, Euler's formula and the classification of surfaces imply that we can identify the homeomorphism type of a closed orientable combinatorial surface by counting its cells.

Finally, we will need the following result, known as the *Jordan curve theorem*. (See Munkres [60, Thm. 36.3].)

Theorem 2.6.11. *Let X be a combinatorial surface homeomorphic to $\mathbf{S}^2$, and let A be a subcomplex of X that is homeomorphic to $\mathbf{S}^1$. Then $X - A$ has two components.* □

2.7 The combinatorial Gauss-Bonnet theorem

In this section, we present a (well-known) combinatorial version of the Gauss-Bonnet theorem from differential geometry, for use in Sect. 6.2. Note that the reader who prefers the Riemann-Hurwitz formula may substitute that instead. For completeness, we include a proof, modelled after the proof in Spivak [82, Chap. 6, Thm. 8] of the differential Gauss-Bonnet theorem.

Definition 2.7.1. We say that a triangulated surface X is a *geometrized polyhedron* if each triangle of X is assigned a nonnegative real number, which we think of as an angle, at each of its vertices. Let v be a vertex of a geometrized polyhedron. We define the *curvature* at v to be the amount by which the sum of the angles meeting at v falls short of 1 rev, and we denote the curvature at a vertex v by $\kappa(v)$.

Definition 2.7.2. We define the integral of the curvature over a triangle σ with angles of A, B, and C revs to be

$$\int_\sigma \kappa = \left(A + B + C - \frac{1}{2} \right) \text{ revs,} \qquad (2.7.1)$$

and we define the integral of the curvature over a geometrized polyhedron X to be

$$\int_X \kappa = \sum_{\sigma \in X} \int_\sigma \kappa + \sum_{v \in X} \kappa(v), \qquad (2.7.2)$$

where σ runs over all triangles in X, and v runs over all 0-cells in X.

Theorem 2.7.3 (Combinatorial Gauss-Bonnet). *If X is a geometrized polyhedron,*

$$\int_X \kappa = \chi(X) \text{ revs,} \qquad (2.7.3)$$

where $\chi(X)$ is the Euler characteristic of X.

Proof. Let V, E, and F be the number of 0-, 1-, and 2-cells of X, respectively. Computing the first term of the right-hand side of (2.7.2), and keeping in mind that $3F = 2E$, since X is a triangulated surface, we have

$$\sum_{\sigma \in X} \int_\sigma \kappa = \sum_{\sigma \in X} \left(\left(\sum_{\text{angles } A \text{ of } \sigma} A \right) - \frac{1}{2} \text{ rev} \right)$$

$$= -\frac{1}{2} F \text{ revs} + \left(\sum_{\text{angles } A \text{ in } X} A \right) \qquad (2.7.4)$$

$$= F \text{ revs} - E \text{ revs} + \left(\sum_{v \in X} (1 \text{ rev} - \kappa(v)) \right)$$

$$= F \text{ revs} - E \text{ revs} + V \text{ revs} - \sum_{v \in X} \kappa(v).$$

Therefore, from Euler's formula (Thm. 2.6.10), we have

$$\int_X \kappa = (F - E + V) \text{ revs} = \chi(X) \text{ revs.} \qquad (2.7.5)$$

$\square$

2.8 Other background material

There are two important background topics that are too complicated to summarize here: finite simple groups and Seifert fibered 3-manifolds.

As mentioned in the introduction, one major motivation for this work is the goal of understanding finite simple groups, especially sporadic groups, and of those, especially the ones related to the Fischer-Griess Monster. For an excellent introduction to the sporadic finite simple groups, see Conway and Sloane [20, Chaps. 10, 11, 29]. More detailed and/or comprehensive treatments may be found in Aschbacher [1] and Griess [37]. For the study of finite groups in general, especially the classification of finite simple groups, see Aschbacher [2], Carter [11], and Gorenstein [34, 35]. See also Gorenstein, Lyons, and Solomon [36], along with the related series. For an authoritative reference on finite simple groups, especially the sporadic finite simple groups, see the ATLAS [16].

As for Seifert fibered 3-manifolds, they will only be considered explicitly in Sects. 7.3 and 13.7, and a lack of familiarity with Seifert fibered 3-manifolds should not pose a problem with the rest of the monograph. (On the other hand, the reader already familiar with Seifert fibered 3-manifolds may have recognized their influence on Sect. 2.3.) For the reader who is interested in learning about Seifert fibered 3-manifolds, we highly recommend Seifert's original paper [79]. More modern treatments may be found in Orlik [69, Chaps. 1, 5] and Scott [78, §1–3].

Part I

The geometry of Seifert group actions

3. Quilts

In this chapter, we introduce quilts, which are diagrams that represent subgroups of Seifert groups. As mentioned in the introduction, we are particularly interested in the Seifert group $\mathbf{B}_3$. We approach the fundamental results on quilts by proving a series of *covering theorems*, starting with the covering theorem for permutation representations (Sect. 3.1), proceeding to the covering theorem for modular quilts (Sect. 3.2), and culminating in the covering theorem for quilts (Sect. 3.3). (Note that Sect. 3.3 contains the principal new material in the chapter.) We end the chapter by discussing some practical aspects of drawing quilts (Sect. 3.4) and collecting assorted remarks and references (Sect. 3.5).

For easier reference, we repeat the following formulas from Chap. 2.

$$(m_1, m_2, n) = \langle V_1, V_2, L \,|\, 1 = V_1^{m_1} = V_2^{m_2} = L^n = V_1 V_2 L \rangle . \tag{2.2.1}$$

$$\left\langle \frac{p_1}{m_1}, \frac{p_2}{m_2}, \frac{q}{n} \right\rangle = \langle Z, V_1, V_2, L \,|\, 1 = [Z, V_1] = [Z, V_2] = [Z, L], \tag{2.3.1}$$
$$Z^{p_1} = V_1^{m_1}, \; Z^{p_2} = V_2^{m_2}, \; Z^q = L^n,$$
$$1 = V_1 V_2 L \rangle .$$

$$\left\langle \frac{p_1}{m_1}, \frac{p_2}{m_2}, 0 \right\rangle = \langle Z, V_1, V_2, L \,|\, 1 = [Z, V_1] = [Z, V_2] = [Z, L], \tag{2.3.2}$$
$$Z^{p_1} = V_1^{m_1}, \; Z^{p_2} = V_2^{m_2},$$
$$1 = V_1 V_2 L \rangle .$$

$$\mathbf{B}_3 \cong \langle Z, V_1, V_2, L \,|\, 1 = [Z, V_1] = [Z, V_2] = [Z, L], \tag{2.4.4}$$
$$Z = V_1^3, \; Z^{-1} = V_2^2,$$
$$1 = V_1 V_2 L \rangle ,$$
$$= \left\langle \frac{1}{3}, -\frac{1}{2}, 0 \right\rangle .$$

Recall also that $\mathbf{PSL}_2(\mathbf{Z}) \cong (3, 2, \infty) \cong \mathbf{B}_3 / \langle Z \rangle$, and that by convention, $R = V_2 V_1$.

3.1 Permutation representations

We review some basic facts about permutation representations, both to establish notation and terminolgy, and also to state an elementary, but perhaps less familiar, result (Thm. 3.1.3).

Recall that if G is group acting transitively on a set Ω, then the stabilizers of points in Ω are precisely a conjugacy class of subgroups of G. On the other hand, a subgroup H of G determines a transitive permutation representation of G on the cosets of H, called the *coset representation induced by H*, and conjugate subgroups determine the same representation. We have:

Theorem 3.1.1. *Let G be a group. Conjugacy classes of subgroups of G correspond bijectively with the representations they induce.* $\square$

From now on we may therefore switch freely between a conjugacy class of subgroups of G and its induced representation.

We also need the naturality of the above correspondence (Thm. 3.1.3). However, to give a precise statement of naturality, we first need to define appropriate morphisms between permutation representations.

Definition 3.1.2. Let G act on the sets Ω_1 and Ω_2. We say that $f : \Omega_1 \to \Omega_2$ is *G-equivariant* if

$$f(x)g = f(xg) \quad \text{for all } x \in \Omega_1, g \in G. \tag{3.1.1}$$

Note that if f is a G-equivariant map from Ω_1 to Ω_2, and G is transitive on Ω_2, then (3.1.1) implies that f is surjective. Therefore, if such a map exists, we say that Ω_1 *covers* Ω_2.

We may now state our naturality result.

Theorem 3.1.3 (Covering theorem). *Let H and K be subgroups of G. H is contained in a conjugate of K if and only if the coset representation induced by H covers the coset representation induced by K.*

Proof. Let Ω_H (resp. Ω_K) be the coset representation induced by H (resp. K). We first show that if Ω_H covers Ω_K, then H is contained in a conjugate of K. Let f be a G-equivariant map from Ω_H to Ω_K. For $h \in H$, h stabilizes the coset H, so by (3.1.1), h also stabilizes $f(H)$. Therefore, considering the action of G on Ω_K, we see that H is contained in the stabilizer of $f(H)$. However, since G is transitive on Ω_K, the stabilizer of $f(H)$ is conjugate to K.

As for the converse, without loss of generality, we may assume that H is a subgroup of K. Define a map f from Ω_H to Ω_K by sending the coset Ha to the coset of K containing Ha. To check the invariance of f, it suffices to show that if $f(Ha) = Kb$, then $f(Hag) = Kbg$ for all $g \in G$. However, by the definition of f, this holds if and only if $Ha \subset Kb$ implies $Hag \subset Kbg$, and this is clear. $\square$

One important aspect of the covering theorem is that the coset representation induced by a *smaller* subgroup covers the one induced by a *larger* subgroup. The same phenomenon can be seen with covering spaces and their fundamental groups, or with Galois extensions and their groups.

Finally, the following definition will be useful later.

Definition 3.1.4. Let $f : \Omega_1 \to \Omega_2$ be a G-equivariant map between transitive permutation representations of G. We define the *degree* of f to be the size of the set $\{x \mid f(x) = y\}$ for any $y \in \Omega_2$.

Note that the degree of f is well-defined, since $f(x) = y$ if and only if $f(x)g = yg$ for any $g \in G$, which holds if and only if $f(xg) = yg$ for any $g \in G$. Note also that f is an isomorphism if and only the degree of f is 1.

3.2 Modular quilts

Let G be the triangle group (m_1, m_2, n) (Sect. 2.2), where n may equal ∞. In this section, we define certain 2-complexes called *modular quilts*, we show that modular quilts are equivalent to transitive permutation representations of G, and we recast the results of the previous section in this context. Note that the material in this section is well-known; in fact, the reader who is familiar with coset diagrams (see App. A) will recognize that modular quilts are essentially coset diagrams for subgroups of (m_1, m_2, n) on the generators V_1 and V_2. See also Sect. 3.5.

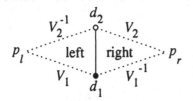
Fig. 3.2.1. A seam

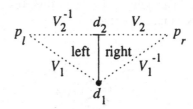
Fig. 3.2.2. A seam, $m_2 = 2$

We begin by describing the basic building block of our 2-complexes. Consider the double triangular region shown in Fig. 3.2.1. Such a region is called a *seam*. More formally, a seam is a 2-complex s constructed as follows.

1. s has four 0-cells, which are marked in Fig. 3.2.1 as d_1 and d_2 (the *dots* of type 1 and 2), p_ℓ, and p_r. By convention, black dots are of type 1, and white dots are of type 2.
2. s has five 1-cells, which consist of one *solid 1-cell*, between d_1 and d_2 (the *midline* of s), and four *dotted 1-cells*, between d_1 and p_ℓ, d_1 and p_r, d_2 and p_ℓ, and d_2 and p_r (marked V_1, V_1^{-1}, V_2^{-1}, and V_2, respectively).

3. s has two 2-cells (its *left* and *right 2-cells*), attached to its 1-skeleton as shown in Fig. 3.2.1.

When $m_2 = 2$, it will be convenient for us to draw dots of type 2 as dashes, instead of white dots, as shown in Fig. 3.2.2. In that case, we call dots of type 2 *dashes* and dots of type 1 simply *dots*.

The diagrams we call *modular quilts* are now made by gluing a set of seams together along their dotted 1-cells in the following manner. We first define the case $n = \infty$ (Defn. 3.2.1), and later proceed to the general case (Defn. 3.2.6).

Definition 3.2.1. A (m_1, m_2, ∞)-*modular quilt* Q is a set of seams with some of their dotted 1-cells identified (in the sense of a quotient space) according to the following rules.

1. For $r = 1, 2$, a dotted 1-cell marked V_r can only be identified with a dotted 1-cell marked V_r^{-1}. Furthermore, all identifications are cellular (n-cells with n-cells), and dots of type r ($r = 1, 2$) may only be identified with other dots of type r.
2. Each dotted 1-cell is identified with precisely one other dotted 1-cell.
3. Q is connected.
4. For $r = 1, 2$, the number of seams around a dot of type r divides m_r.

When the group (m_1, m_2, ∞) is understood, we simply call Q a *modular quilt*.

For $r = 1, 2$, the meeting of m_r seams around a dot of type r is called a *vertex of type r*. Similarly, the meeting of fewer than m_r seams around a dot of type r is called a *collapsed vertex of type r*. The *collapsing index* of a modular quilt at a collapsed vertex of type r is m_r divided by the number of seams meeting there.

Fig. 3.2.3. A vertex in a
$(3, 2, \infty)$-quilt

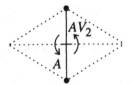

Fig. 3.2.4. An edge in a
$(3, 2, \infty)$-quilt

Again, our terminology differs in the case $m_2 = 2$. In that case, a (collapsed) vertex of type 2 is called a (collapsed) *edge*, and a (collapsed) vertex of type 1 is just called a (collapsed) *vertex*. In particular, in a $(3, 2, \infty)$-modular quilt, a vertex consists of 3 seams meeting at their dots (Fig. 3.2.3), and an edge consists of 2 seams meeting at their dashes (Fig. 3.2.4); and similarly,

Fig. 3.2.5. A collapsed vertex in a $(3, 2, \infty)$-quilt

Fig. 3.2.6. A collapsed edge in a $(3, 2, \infty)$-quilt

a collapsed vertex is a single seam identified with itself around its dot (Fig. 3.2.5), and a collapsed edge is a single seam identified with itself around its dash (Fig. 3.2.6), with collapsing indices 3 and 2, respectively. (The solid 1-cells and the dash in Fig. 3.2.4 also show why we use the slightly confusing term of "edge" to refer to vertices of type 2 when $m_2 = 2$.)

To complete our picture of what a modular quilt looks like, and to extend our definition to the case $n < \infty$, we next need the following definition.

Definition 3.2.2. Let Q be a (m_1, m_2, ∞)-modular quilt, and let X be the subcomplex of Q consisting of the union of the midlines and dots (of both types) of Q. A connected component of $Q - X$ is called a *patch*.

Remark 3.2.3. As can be seen in larger examples (see, for instance, Chap. 11), modular quilts look like patches sewn together at seams, which is why we call them "quilts."

Now, as shown in Fig. 3.2.7, since left and right 2-cells must alternate as we "walk around" a patch, every patch contains either an even number or an infinite number of 2-cells. Furthermore, every patch contains a single 0-cell, which we call the *patch point* of the patch.

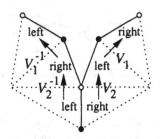

Fig. 3.2.7. Walking around patches

Note also that if a patch contains an infinite number of 2-cells, then its patch point has no neighborhood with a compact closure, and therefore cannot be a point in a surface. Therefore, by convention, we topologize modular quilts by deleting such infinite patch points. Given this convention, it is not hard to see that:

Proposition 3.2.4. *A modular quilt is an orientable surface, and is compact if and only if it is made from a finite number of seams.*   □

In any case, since every patch contains either an even or an infinite number of 2-cells, the following definition makes sense.

Definition 3.2.5. Let P be a patch of a modular quilt Q. If P contains $2n$ 2-cells, we say that P has *order* n. (Note that n may equal ∞.)

We can now finally give the general definition of modular quilt.

Definition 3.2.6. A (m_1, m_2, n)-*modular quilt* $(n < \infty)$ is an (m_1, m_2, ∞)-modular quilt Q such that the order of every patch of Q divides n. As before, if the group (m_1, m_2, n) is understood, Q is just called a *modular quilt.*

It will also later be convenient to have the following definitions.

Definition 3.2.7. Let Q be an (m_1, m_2, n)-modular quilt, with $n < \infty$. The *collapsing index* of a patch of Q is defined to be n divided by the order of the patch. A patch of collapsing index > 1 is said to be *collapsed*, and a patch of collapsing index 1 is said to be *uncollapsed.*

Note that we do not define collapsed patches in (m_1, m_2, ∞)-modular quilts.

Definition 3.2.8. The *global collapsing index* of an (m_1, m_2, n)-modular quilt Q is defined to be the least common multiple of all of the collapsing indices of all vertices (of both types) and patches of Q. Similarly, the *global collapsing index* of an (m_1, m_2, ∞)-modular quilt Q is defined to be the least common multiple of all of the collapsing indices of all vertices (of both types) of Q.

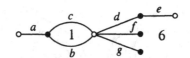

Fig. 3.2.8. A $(6, 5, \infty)$-modular quilt

Example 3.2.9. Figure 3.2.8 shows a $(6, 5, \infty)$-modular quilt Q, drawn using two conventions: First, we omit the dotted 1-seams of Q for clarity; and second, since Q happens to have genus 0, we draw Q by omitting a point "at infinity", or in other words, by thinking of Q as a planar diagram. Let a, b, c, d, e, f, g be the seams whose midlines are marked as shown in Fig. 3.2.8. The reader may wish to verify, as an exercise in unwrapping the definitions given so far, that Q contains:

- Four collapsed vertices of type 1, of collapsing index 2, 3, 6, and 6, consisting of the seams $\{a, b, c\}$, $\{d, e\}$, $\{f\}$, and $\{g\}$, respectively;

- One uncollapsed vertex of type 2, consisting of the seams $\{b, g, f, d, c\}$;
- Two collapsed vertices of type 2, both of collapsing index 5, consisting of the seams $\{a\}$ and $\{e\}$, respectively; and
- Two patches, one of order 1, and one of order 6 (both marked on the diagram).

As a $(6, 5, \infty)$-modular quilt, Q therefore has global collapsing index 30.

Note that Q is also a $(6, 5, 6n)$-modular quilt for any $n \in \mathbf{Z}$. As a $(6, 5, 6n)$-modular quilt, the patches of Q have collapsing index n and $6n$, so as a $(6, 5, 6n)$-quilt, Q has global collapsing index $\mathrm{lcm}(6n, 30)$.

Having established the basic definitions, we next show that a (m_1, m_2, n)-modular quilt defines a permutation representation of (m_1, m_2, n) in the following manner.

Definition 3.2.10. Let Q be a (m_1, m_2, n)-modular quilt. The *modular quilt representation* of (m_1, m_2, n) induced by Q is a permutation representation of (m_1, m_2, n) defined by the following rules:

1. The domain of the permutation representation is the set of seams of Q.
2. For $r = 1, 2$, if the V_r dotted 1-cell of the seam A is identified with the V_r^{-1} dotted 1-cell of the seam B, then $AV_r = B$.

Note that since $L = V_2^{-1}V_1^{-1}$, these rules define an action of each generator of (m_1, m_2, n).

Example 3.2.11. Consider the case of $(m_1, m_2, n) = (3, 2, \infty)$. In that case, the rules in Defn. 3.2.10 are described by Figs. 3.2.3–3.2.6, pp. 32–33, in which each seam is marked with its corresponding permuted object, and the actions of V_1 and V_2 are indicated by the circular arrows. Note that, as dictated by the geometry we have chosen, we think of the actions around either an edge or a vertex as going counterclockwise.

Note also that since $L = V_2^{-1}V_1^{-1}$, the action of L is given by a "left turn" around a patch, starting from a left 2-cell, as indicated in Fig. 3.2.7, p. 33. Similarly, starting from a right 2-cell, we see that $R = V_2V_1$ is given by a "right turn" around a patch.

Example 3.2.12. More concretely, let Q be the $(6, 5, \infty)$-modular quilt shown in Fig. 3.2.8. The modular quilt representation induced by Q is the permutation representation ρ of $(6, 5, \infty)$ given by

$$\rho(V_1) = (a\ b\ c)(d\ e)(f)(g), \qquad \rho(V_2) = (a)(b\ g\ f\ d\ c)(e),$$
$$\rho(L) = (a\ c\ e\ d\ f\ g)(b), \qquad \rho(R) = (a\ b\ g\ f\ e\ d)(c), \tag{3.2.1}$$

where a, b, c, d, e, f, g are the seams whose midlines are marked as shown in Fig. 3.2.8.

We may also reverse the above procedure to obtain a modular quilt from a permutation represention.

Definition 3.2.13. Suppose that (m_1, m_2, n) acts transitively on a set Ω. Let Q be the following 2-complex:

1. The seams of Q correspond bijectively with elements of Ω.
2. For $r = 1, 2$ and all $A \in \Omega$, the V_r dotted 1-cell of seam A is identified with the V_r^{-1} dotted 1-cell of seam B, in the manner described by Defn. 3.2.1 (1), if and only if $AV_r = B$.

We call Q the modular quilt of Ω. Similarly, we define the modular quilt of $H \leq (m_1, m_2, n)$ to be the modular quilt of the representation induced by H.

Of course, we must check that Defns. 3.2.10 and 3.2.13 actually produce permutation representations and modular quilts, respectively, which we do in the following theorem.

Theorem 3.2.14. *The operations described by Defns. 3.2.10 and 3.2.13 produce a bijection between (m_1, m_2, n)-modular quilts and transitive permutation representations of (m_1, m_2, n).*

Proof. We first check that if Q is a (m_1, m_2, n)-modular quilt, then Defn. 3.2.10 defines a transitive permutation representation of (m_1, m_2, n). Transitivity follows from the connectedness of Q, so we merely have to check that the rules in Defn. 3.2.10 respect the relations $1 = V_1^{m_1}$, $1 = V_2^{m_2}$, and $1 = L^n$ (when $n < \infty$). The first two relations follow directly from condition (4) of Defn. 3.2.1. As for $1 = L^n$ ($n < \infty$), since the order of every patch divides n, if we make n "left turns" from a given starting point, we return to that starting point. (See Fig. 3.2.7, p. 33.) Therefore, the relation is respected.

Conversely, if Ω is a transitive permutation representation of (m_1, m_2, n) and Q is the 2-complex described by Defn. 3.2.13, then it is clear from the nature of permutation representations that Q satisfies the first two conditions of Defn. 3.2.1. The transitivity of Ω implies that Q is connected, so it remains only to check the vertex and patch conditions. However, since Ω respects the relations $1 = V_1^{m_1} = V_2^{m_2}$, Q must satisfy condition (4) of Defn. 3.2.1. Furthermore, when $n < \infty$, since $L = V_2^{-1}V_1^{-1}$ and $1 = L^n$, running the "left turn" argument in reverse, we see that the order of any patch of Q must divide n.

Finally, we leave it as an exercise for the reader to verify that the operations defined by Defns. 3.2.10 and 3.2.13 are inverses, and therefore, both bijections.   □

Remark 3.2.15. Note that for $m_1 = 3$, $m_2 = 2$, Thm. 3.2.14 says that modular quilts are equivalent to transitive permutation representations of the modular group $\mathbf{PSL_2(Z)}$, which is why we call these quilts "modular". See also the beginning of Sect. 3.5.

Note that under the correspondence of Thm. 3.2.14, (collapsed) vertices of type r correspond with V_r-orbits. Similarly, patches correspond with the following objects.

Definition 3.2.16. Let Ω be a permutation representation of (m_1, m_2, n). Note that if A is an L-orbit of Ω, then since

$$R = V_2 V_1 = (V_1^{-1} L^{-1} V_1) = (V_2 L^{-1} V_2^{-1}), \qquad (3.2.2)$$

$AV_1 = AV_2^{-1}$ is an R-orbit of Ω. We therefore define an L/R-*orbit* of Ω to be an ordered pair (A, B), where A is an L-orbit of Ω, and $B = AV_1 = AV_2^{-1}$ is the associated R-orbit.

The "left turn" argument from the proof of Thm. 3.2.14 and its corresponding "right turn" analogue then immediately give the following corollary.

Corollary 3.2.17. *Let Ω be the representation induced by a modular quilt Q, and let P be a patch of Q. Then the objects of Ω associated with the seams whose left (resp. right) 2-cells are in P form an L-orbit (resp. R-orbit) of Ω. In other words, the patches of Q correspond bijectively with the L/R-orbits of Ω.* □

Finally, we prove a version of the covering theorem (Thm. 3.1.3) for modular quilts. To state this theorem, however, we first need to define suitable morphisms.

Definition 3.2.18. Let Q_1 and Q_2 be modular quilts. A continuous cellular map from Q_1 to Q_2 that sends seams to seams, dots of type r to dots of type r ($r = 1, 2$), and left (resp. right) 2-cells to left (resp. right) 2-cells is called a *modular quilt morphism*.

The covering theorem and Thm. 3.2.14 then imply the following theorem.

Theorem 3.2.19 (Covering theorem for modular quilts). *Let H and K be subgroups of (m_1, m_2, n). Then H is contained in a conjugate of K if and only if there is a morphism from the modular quilt of H to the modular quilt of K.*

Proof. Let Q_H (resp. Q_K) be the modular quilt of H (resp. K), and let f be a map from the seams of Q_H to the seams of Q_K. From Thm. 3.1.3, it suffices to show that f defines a modular morphism if and only if f is (m_1, m_2, n)-equivariant. So suppose f is a modular morphism. In that case, since f is continuous, if, for $r = 1, 2$, the V_r dotted 1-cell of A is identified with the V_r^{-1} dotted 1-cell of B in Q_H, then the V_r dotted 1-cell of $f(A)$ is identified with the V_r^{-1} dotted 1-cell of $f(B)$ in Q_K. From the definition of the modular quilt representation, it follows that

$$f(A)V_r = f(B) = f(AV_r) \qquad (3.2.3)$$

for $r = 1, 2$ and all seams A of Q_H. In other words, f is equivariant. The same argument backwards yields the converse, and the theorem follows. □

It follows that modular morphisms are always surjective. Therefore, we also call them *modular coverings*.

Remark 3.2.20. Note that the above material is essentially equivalent to the topological version of Riemann's existence theorem for covers of the sphere with specified ramification at 3 points. Compare, for instance, Völklein [86, Chap. 4].

3.3 Quilt diagrams and quilts

Our goal now is to lift the results of the previous section to subgroups of Seifert groups (Sect. 2.3). After some prelimaries, we begin by defining *quilt diagrams*, which represent subgroups of Seifert groups of the form $\left\langle \frac{p_1}{m_1}, \frac{p_2}{m_2}, 0 \right\rangle$. We then proceed as in Sect. 3.2, obtaining a bijection theorem (Thm. 3.3.20), a covering theorem (Thm. 3.3.25), and an interpretation of patches (Thm. 3.3.28). The key point is that transitive permutation representations of a Seifert group correspond not to a single diagram but an *equivalence class* of diagrams called a *quilt* (Defn. 3.3.19). Finally, we discuss quilts that represent subgroups of $\left\langle \frac{p_1}{m_1}, \frac{p_2}{m_2}, \frac{q}{n} \right\rangle$ for $n \neq \infty$, and we prove a "pullback" theorem for later use.

Notation 3.3.1. Throughout this section, if $\Sigma = \left\langle \frac{p_1}{m_1}, \frac{p_2}{m_2}, \frac{q}{n} \right\rangle$ is a Seifert group, we let $\overline{\Sigma}$ denote (m_1, m_2, n) (its reduction mod $\langle Z \rangle$).

For the moment, consider the general case of a Seifert group $\Sigma = \left\langle \frac{p_1}{m_1}, \frac{p_2}{m_2}, \frac{q}{n} \right\rangle$. Quite naturally, the kernel $\langle Z \rangle$ of the projection from Σ to $\overline{\Sigma}$ is the key to lifting modular quilts to quilts. We therefore begin with an observation about $\langle Z \rangle$ that leads to a definition.

Proposition 3.3.2. *If Σ acts transitively on Ω, then all elements of Ω have the same stabilizer in $\langle Z \rangle$.*

Proof. For any $x \in \Omega$ and $\pi \in \Sigma$, $x = xZ^i$ if and only if $x\pi = xZ^i\pi = x\pi Z^i$, since Z is central. $\qquad\square$

Definition 3.3.3. Let Σ act transitively on a set Ω. We define the *modulus* of Ω, or equivalently, of any stabilizer Δ of a point in Ω, to be the unique nonnegative M such that $\langle Z \rangle \cap \Delta = \langle Z^M \rangle$. In other words, if the Z-orbits of Ω are finite, M is the size of a Z-orbit, and if the Z-orbits of Ω are infinite, then $M = 0$.

From now until Defn. 3.3.31, we restrict our attention to Seifert groups of the form $\Sigma = \left\langle \frac{p_1}{m_1}, \frac{p_2}{m_2}, 0 \right\rangle$. We begin to define quilts of subgroups of these groups with the following.

Definition 3.3.4. Let M be a nonnegative integer, and let Q be a $\overline{\Sigma}$-modular quilt whose dotted 1-cells are labelled with integers mod M. We define the *inflow* of a vertex v of Q (of type 1 or 2) to be the sum of the numbers on the dotted 1-cells touching v. We also sometimes call the inflow of v the *flow into* v.

We say that the pair (Q, M), including the labelling of the dotted 1-cells of Q, is a Σ-*quilt diagram* if every vertex of type r ($r = 1, 2$), collapsing index i (Defn. 3.2.1), and inflow t satisfies the *flow rule*

$$ti \equiv p_r \quad (\text{mod } M). \tag{3.3.1}$$

Note that if i is relatively prime to M, then p_r/i is well-defined (mod M), and the flow rule becomes

$$t \equiv \frac{p_r}{i} \quad (\text{mod } M). \tag{3.3.2}$$

As with modular quilts, if the group Σ is understood, then (Q, M) is just called a *quilt diagram*. The modular quilt Q, without its labels, is called the *modular structure* of (Q, M), and M is called the *modulus* of (Q, M). By abuse of notation, we sometimes refer to a quilt diagram (Q, M) by its underlying labelled modular structure Q.

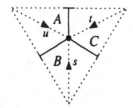
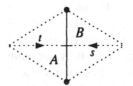

Fig. 3.3.1. Quilt diagram labels as arrow flows, type 1

Fig. 3.3.2. Quilt diagram labels as arrow flows, type 2

Remark 3.3.5. Geometrically, the $\mathbf{Z}/M$ labels of Defn. 3.3.4 are perhaps best thought of as *arrow flows*. For instance, the flow into the vertex of type 1 in Fig. 3.3.1 is $s + t + u$, and the flow into the vertex of type 2 in Fig. 3.3.2 is $s + t$. (Note that we use the drawing conventions of the case where $m_2 = 2$.) Considering the numbers in a quilt diagram to be arrow flows is useful because an arrow flow of $-t$ can be written as a flow of t with the arrow reversed, a convention we use freely in the sequel.

In other words, for the reader familiar with elementary homology theory (see Sect. 2.5), when Q is finite, the $\mathbf{Z}/M$ labels are really just a 1-chain on the dotted 1-cells of Q, with all 1-cells oriented towards the vertices of type 1 and 2 of Q.

Now, if (Q, M) is a Σ-quilt diagram, then for every vertex of Q of type r ($r = 1, 2$) and collapsing index i, (3.3.1) and elementary facts about congruences imply that $\gcd(M, i)$ divides p_r. Therefore, since the collapsing index of a vertex of Q is simply a function of the modular quilt structure of Q, the following definition makes sense.

Definition 3.3.6. Let Σ be a Seifert group, let Q be a $\overline{\Sigma}$-modular quilt, and let M be a nonnegative integer. If, for every vertex of Q of type r and collapsing index i, $\gcd(M, i)$ divides p_r, then M is said to be *compatible* with Q and Σ.

Now, suppose that Σ is geometric (Defn. 2.3.3), as it is in the cases in which we are most interested. In that case, since the collapsing index i at a vertex of type r is always a divisor of m_r, a modulus M is compatible with a modular quilt Q if and only if $\gcd(M, i) = 1$ for all collapsing indices i of all vertices in Q. Furthermore, as noted in Defn. 3.3.4, in a Σ-quilt diagram (Q, M), the flow rule assumes the simpler form (3.3.2). In other words, recalling Defn. 3.2.8, we have:

Proposition 3.3.7. *Let* $\Sigma = \left\langle \frac{p_1}{m_1}, \frac{p_2}{m_2}, 0 \right\rangle$ *be a geometric Seifert group, and let Q be a $\overline{\Sigma}$-modular quilt. A nonnegative integer M is compatible with Q and Σ if and only if it is relatively prime to the global collapsing index of Q. Furthermore, (3.3.2) holds at every vertex of Q.* $\square$

Example 3.3.8. To make the above definitions a little more concrete, in this example, let $\Sigma = \mathbf{B}_3 = \langle \frac{1}{3}, -\frac{1}{2}, 0 \rangle$ and $\overline{\Sigma} = \mathbf{PSL}_2(\mathbf{Z}) = (3, 2, \infty)$. Since Σ is geometric, the notion of compatibile modulus (Defn. 3.3.6) can be understood from the simpler formulation of Prop. 3.3.7. That is, any modulus is compatible with a modular quilt with no collapsed vertices or edges; any odd modulus is compatible with a modular quilt with collapsed edges only; any modulus not divisible by 3 is compatible with a modular quilt with collapsed vertices only; and any modulus relatively prime to 6 is compatible with a modular quilt with both collapsed edges and collapsed vertices.

Recall that since $m_2 = 2$, vertices of type 2 are called edges and vertices of type 1 are simply called vertices. In these terms, for a $\mathbf{B}_3$-quilt diagram (Q, M), the flow rule of Defn. 3.3.4 becomes:

1. If t is the inflow of a vertex of collapsing index i, then $ti \equiv 1 \pmod{M}$. Note that i is always either 1 (an uncollapsed vertex) or 3 (a collapsed vertex).
2. If t is the inflow of an edge of collapsing index i, then $ti \equiv -1 \pmod{M}$. Note that i is always either 1 or 2.

In other words:

1. The flow into a vertex is 1 mod M.
2. The flow into an edge is -1 mod M. In other words, the flow out of an edge is 1 mod M.
3. The flow into a collapsed vertex is $\frac{1}{3}$ mod M.
4. The flow into a collapsed edge is $-\frac{1}{2}$ mod M. In other words, the flow out of a collapsed edge is $\frac{1}{2}$ mod M.

Note that since M is compatible with Q, $\frac{1}{3}$ and $\frac{1}{2}$ mod M make sense if the appropriate collapse appears.

Exercise 3.3.9. Check that the annotated 2-complex in Fig. 3.3.3, with 3 seams and modulus $M = 5$, is a $\mathbf{B}_3$-quilt diagram. Note that we use our arrow flow direction convention freely. In addition, an arrow flow of 1 is indicated by an unmarked arrow, and an arrow flow of 0 is indicated by omitting the arrow completely.

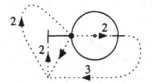

Fig. 3.3.3. A $\mathbf{B}_3$-quilt diagram, $M = 5$

Now, still fixing a Seifert group Σ with $n = \infty$, and considering a Σ-quilt diagram (Q, M), we can construct the following permutation representation of Σ.

Definition 3.3.10. Let (Q, M) be a Σ-quilt diagram. We define a (putative) permutation representation of Σ by the following rules:

1. The domain of the permutation representation is the set of pairs (A, j), where A is a seam of Q, and j is an integer mod M.
2. The action of Z is defined by

$$(A, j)Z^\ell = (A, j + \ell). \tag{3.3.3}$$

Note that each seam therefore determines a unique Z-orbit.
3. For $r = 1, 2$, the action of V_r is defined as follows. If the V_r dotted 1-cell of seam A is identified with the V_r^{-1} dotted 1-cell of seam B, and the identified 1-cell is labelled with t, as shown in Fig. 3.3.4 (for a vertex of type 1), then we define

$$(A, j)V_r = (B, j + t). \tag{3.3.4}$$

(Note that since $V_1 V_2 L = 1$, we do not need to define the action of L separately.)

We call this representation the *quilt representation* of Σ induced by (Q, M), or simply the representation induced by (Q, M).

As in Sect. 3.2, we must check that this actually defines a representation of Σ, which we do in the following proposition.

Fig. 3.3.4. Defining the quilt representation

Proposition 3.3.11. *Let (Q, M) be a quilt diagram. The representation induced by (Q, M) is a permutation representation of Σ with modulus M. Furthermore, the action of Σ on the Z-orbits of the quilt representation naturally reduces to the representation of $\overline{\Sigma}$ induced by the modular quilt Q.*

Proof. Examining presentation (2.3.2) (see the beginning of this chapter), and keeping in mind that the generator L is completely redundant, we see that for the quilt representation to be a valid representation of Σ, it needs only to respect the relations $1 = [Z, V_r]$ and $Z^{p_r} = V_1^{m_r}$ for $r = 1, 2$. The commutator relations follow easily from (3.3.3) and (3.3.4), so it remains to check that if A is a seam of Q, then $(A, j)V_r^{m_1} = (A, j)Z^{p_r}$ for $r = 1, 2$ and all j (mod M).

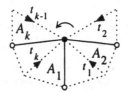

Fig. 3.3.5. Flows around a vertex (type 1)

Now, consider a vertex of Q of type r ($r = 1, 2$) and collapsing index i that contains the seams $A = A_1, A_2, \ldots, A_k$ in counterclockwise order, as shown in Fig. 3.3.5. Let $t = t_1 + \cdots + t_k$ be the total flow into the vertex. Working around the vertex, we obtain:

$$(A_1, j)V_r^k = (A_2, j + t_1)V_r^{k-1}$$

$$\vdots \tag{3.3.5}$$

$$= (A_k, j + t_1 + \cdots + t_{k-1})V_r = (A_1, j + t).$$

However, since the definition of collapsing index (Defn. 3.2.1) implies that $ki = m_r$, and the definition of quilt diagram (Defn. 3.3.4) implies that $ti \equiv p_r$ (mod M), we see that

$$(A, j)V_r^{m_r} = (A, j)(V_r^k)^i = (A, j + ti) = (A, j + p_r) = (A, j)Z^{p_r}. \tag{3.3.6}$$

Then, since a Z-orbit of the quilt representation is precisely the set of all (A, j) for a given seam A of Q, we also see that the modulus of the representation is precisely M, and the first statement of the proposition follows.

As for the second statement, we observe that the action of Σ on the first factor of (A, j) contains $\langle Z \rangle$ in its kernel, and reduces mod $\langle Z \rangle$ to the definition of the modular quilt representation. (Compare Defns. 3.2.10 and 3.3.10.) The proposition follows. $\square$

Definition 3.3.12. We define the *order* of a quilt diagram to be the number of objects in the representation it induces. In other words, the order of (Q, M) is either M times the number of seams in Q, for $M > 0$, or infinity, for $M = 0$.

Example 3.3.13. Let $\Sigma = \langle \frac{1}{3}, -\frac{1}{2}, 0 \rangle \cong \mathbf{B}_3$, and consider the quilt diagram Q with modulus $M = 5$ from Fig. 3.3.3, p. 41. Since Q has 3 seams, (Q, M) induces a permutation representation of Σ on $3 \cdot 5 = 15$ objects, which means that (Q, M) has order 15. In terms of subgroups, the representation induced by Q therefore corresponds to a certain conjugacy class of subgroups of Σ of index 15 that projects down to a conjugacy class of subgroups of $\overline{\Sigma} \cong \mathbf{PSL}_2(\mathbf{Z})$ of index 3. (For readers familiar with modular subgroups, the latter conjugacy class is the conjugacy class of $\Gamma_0(2)$.)

Exercise 3.3.14. Find permutations for the actions of V_1 and V_2 in the representation induced by the quilt diagram in Fig. 3.3.3, p. 41, and check that the permutations satisfy $Z = V_1^3 = V_2^{-2}$. (Hint: Mod $\langle Z \rangle$, the permutations have the form $V_1 \mapsto (A\ B\ C)$ and $V_2 \mapsto (A)(B\ C)$.)

Continuing as before, we would now like to define the quilt diagram of a transitive permutation representation of Σ. However, as we shall see, each such representation usually corresponds to many different quilt diagrams, so instead, we have the following definition.

Definition 3.3.15. Let Σ act transitively on the set Ω, with modulus M, and let $\overline{\Omega}$ be the set of Z-orbits of Ω. Note that the action of Σ on Ω defines an action of $\overline{\Sigma}$ on $\overline{\Omega}$. Let Q be the $\overline{\Sigma}$-modular quilt of $\overline{\Omega}$ (Defn. 3.2.13). We have the following procedure for labelling the dotted 1-seams of Q with integers mod M.

1. Identify each seam of Q with the unique Z-orbit to which it corresponds.
2. In each Z-orbit A, choose a representative $(A, 0)$, and define $(A, j) = (A, 0)Z^j$ for any j (mod M).
3. For $r = 1, 2$, if the seams/Z-orbits $A, B \in \overline{\Omega}$ satisfy $AV_r = B$, then $(A, 0)V_r = (B, t)$ for a unique t (mod M). In that case, we label the V_r dotted 1-cells of A (or equivalently, the V_r^{-1} dotted 1-cell of B) with t, as shown in Fig. 3.3.4, p. 42.

When Q is given this labelling, we call (Q, M) the quilt diagram of the action of Σ on Ω with respect to the representatives $(A, 0)$ $(A \in \overline{\Omega})$.

Proposition 3.3.16. *Let Σ act transitively on the set Ω, with modulus M. The annotated modular quilt obtained from any choice of representatives $(A, 0)$ $(A \in \overline{\Omega})$ is a quilt diagram.*

Proof. From Defn. 3.3.4, we see that all we have to check is the flow rule (3.3.1). However, this essentially follows from the proof of Prop. 3.3.11, run backwards. For instance, suppose $t = t_1 + \cdots + t_k$ is the total flow into the vertex of type r $(r = 1, 2)$ shown in Fig. 3.3.5, p. 42, and i is the collapsing index of the vertex. Since, by definition, $ki = m_r$ and $V_r^{m_r} = Z^{p_r}$, applying (3.3.5) in reverse, we get

$$(A, 0)Z^{p_r} = (A, 0)V_r^{m_r} = (A, 0)(V_r^k)^i = (A, ti) = (A, 0)Z^{ti}. \qquad (3.3.7)$$

However, since the stabilizer of $(A, 0)$ in $\langle Z \rangle$ is generated by Z^M, we must have $ti \equiv p_r \pmod{M}$. The proposition follows. $\qquad \square$

To describe precisely how the quilt diagram of a transitive permutation representation of Σ depends on the choice of representatives of Z-orbits, we need the following definition.

Definition 3.3.17. Let A be the seam shown in Fig. 3.3.6. Recalling our conventions about directions of arrow flows, we call the flow shown in Fig. 3.3.6 a *boundary flow* of $+j$ around A.

Fig. 3.3.6. Boundary flow of $+j$

Fig. 3.3.7. Self-touching boundary flow of $+j$

Note that if A is attached to itself (for instance, if the V_1 dotted line of A is attached to the V_1^{-1} dotted line of A), then the corresponding added boundary flows cancel, and there is no effect on that part of the quilt diagram, as shown in Fig. 3.3.7.

Proposition 3.3.18. *Replacing the Z-orbit representative $(A, 0)$ with the representative $(A, j) = (A, 0)Z^j$ has the effect of adding a boundary flow of $+j$ to the dotted 1-cells around A in the associated quilt diagram.*

Proof. Clearly, only the dotted 1-cells touching A are affected. Therefore, suppose for $r = 1, 2$, the seam A is attached to the seam B at its V_r 1-cell and the seam C at its V_r^{-1} 1-cell, as shown in Fig. 3.3.8. Then by Defn. 3.3.15,

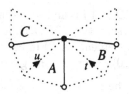

Fig. 3.3.8. Seams touching A

$$(A,0)V_r = (B,t),$$
$$(C,0)V_r = (A,u). \tag{3.3.8}$$

Therefore, since

$$(A,j)V_r = (B,t+j),$$
$$(C,0)V_r = (A,j+(u-j)), \tag{3.3.9}$$

the effect of changing $(A,0)$ to (A,j) is that we have to change t to $t+j$ and u to $u-j$, which is precisely what happens when we add a boundary flow of $+j$. The case where A touches itself is similar, and the proposition follows. $\qquad\square$

Our use of the term "boundary" (compare Figs. 2.5.5 and 2.5.6, p. 21) is meant to suggest the following definition.

Definition 3.3.19. Let Q be a Σ-quilt diagram. Since every arrow flow is only affected by at most two seams, adding a boundary flow around *every* seam of Q is well-defined, even if Q has infinitely many seams. We call a (possibly infinite) sum of boundary flows around seams simply a *boundary*. Note that this sum may be infinite, as long as every seam only appears finitely many times in the sum.

We say that two Σ-quilt diagrams Q_1 and Q_2 are *homologous* if Q_1 and Q_2 have the same modular structure and their arrow flows differ by a boundary. (Note that if Q_1 and Q_2 have only finitely many seams, then this is just the usual definition of homologous; compare Sect. 2.5.) A homology class of Σ-quilt diagrams is called a *Σ-quilt*. As usual, if Σ is understood, then a Σ-quilt is just called a *quilt*.

The following is then an immediate consequence of Props. 3.3.11, 3.3.16, and 3.3.18.

Theorem 3.3.20 (Bijection theorem). *Let Σ be a Seifert group of the form $\left\langle \frac{p_1}{m_1}, \frac{p_2}{m_2}, 0 \right\rangle$. The operations from Defns. 3.3.10 and 3.3.15 give a bijection between Σ-quilts and transitive permutation representations of Σ.* $\quad\square$

Notation 3.3.21. We set the following conventions for quilt diagrams and the representations they induce. If (Q, M) is a quilt diagram, then the domain of the corresponding transitive permutation representation is the set of all

(A, j), where A is a seam of Q and j is an integer mod M. Conversely, if Ω is a transitive permutation representation of Σ, a choice of Z-orbit representatives is given by the choice of an element $(A, 0)$ in each Z-orbit/seam A. In either case, we define $(A, j) = (A, 0)Z^j$ for any integer j mod M.

Definition 3.3.22. Let Ω be a transitive permutation representation of a Seifert group Σ, and let $\Delta \leq \Sigma$ be the stabilizer of some $x \in \Omega$. Because of Thm. 3.3.20, we may define the *quilt of* Ω (resp. the *quilt of* Δ), denoted by $Q(\Omega)$ (resp. $Q(\Delta)$), to be the quilt corresponding with Ω in Thm. 3.3.20.

Remark 3.3.23 (Homology vs. cohomology). The reader may object to our use of homological terms to describe quilt flows, when they actually behave cohomologically. For instance, quilt flows naturally *pull back* instead of pushing forward. (See, for example, Thm. 3.3.34.) We have chosen to use this slight misnomer because it is the natural terminology with respect to the way we have chosen to draw the diagrams, and because it makes the boundary rule easier to remember. However, we warn the reader that quilt flows really are cohomology, or dually, "homology with infinite chains" (see Munkres [60, §5]). See also the end of Sect. 6.1.

Our next goal in this section is to prove the covering theorem for quilts (Thm. 3.3.25). Again, we first define an appropriate notion of morphism.

Definition 3.3.24. Let (Q_1, M_1) and (Q_2, M_2) be quilt diagrams. We say that a modular morphism $q : Q_1 \to Q_2$ is a *quilt morphism* from (Q_1, M_1) to (Q_2, M_2) if the following conditions are satisfied.

1. M_2 divides M_1.
2. For every dotted 1-cell d in Q_1, if d is labelled with the arrow flow t, and $q(d)$ is labelled with u, then $t \equiv u \pmod{M_2}$. In other words, q sends the flows of Q_1 to the flows of Q_2, mod M_2.

If Q_1 and Q_2 are quilts, we say that a modular morphism $q : Q_1 \to Q_2$ is a *quilt morphism* if there is some quilt diagram Q_1' in the class of Q_1 and some quilt diagram Q_2' in the class of Q_2 such that q is a quilt morphism from Q_1' to Q_2'. We also say that Q_1 *covers* Q_2.

Theorem 3.3.25 (Covering theorem for quilts). *Let Δ_1 and Δ_2 be subgroups of Σ. Δ_1 is a subgroup of a conjugate of Δ_2 if and only if $Q(\Delta_1)$ covers $Q(\Delta_2)$.*

Proof. From the covering theorem for permutation representations, it is enough to show that if Σ acts transitively on Ω_1 and Ω_2, then Ω_1 covers Ω_2 if and only if $Q(\Omega_1)$ covers $Q(\Omega_2)$. For $i = 1, 2$, let $\overline{\Omega}_i$ be the Z-orbits of Ω_i, and let (Q_i, M_i) be the quilt of Ω_i.

We first show that if Ω_1 covers Ω_2, then Q_1 covers Q_2. Let f be a Σ-equivariant map from Ω_1 to Ω_2. The equivariance of f implies that f is surjective, that f preserves Z-orbits, and that M_2 divides M_1. Choose Z-orbit representatives $(B, 0)$ for all $B \in \overline{\Omega}_2$. Next, lift these representatives to

Ω_1; that is, for all $A \in \overline{\Omega}_1$, choose Z-orbit representatives $(A, 0)$ such that if $f(A) = B$, then $f((A, 0)) = (B, 0)$. Finally, use these Z-orbit representatives to specify quilt diagram representatives Q_1 and Q_2. Now, from the covering theorem for modular quilts, f induces a modular morphism $q : Q_1 \to Q_2$, so it remains to check the flow condition of Defn. 3.3.24. However, if, for $r = 1, 2$ and $A, C \in \overline{\Omega}_1$, we have $(A, 0)V_r = (C, t)$, then

$$f((A, 0))V_r = f((A, 0)V_r) = f((C, t)) = f((C, 0)Z^t) = f((C, 0))Z^t. \quad (3.3.10)$$

Therefore, if the dotted 1-cell joining A and C is labelled t (mod M_1), then the dotted 1-cell joining $q(A)$ and $q(C)$ is labelled t (mod M_2). In other words, q sends the flows of Q_1 to Q_2 (mod M_2).

Conversely, suppose there exist quilt diagram representatives Q_1 and Q_2 such that there exists a quilt morphism $q : Q_1 \to Q_2$. Choose Z-orbit representatives $(A, 0)$ for Ω_1 and Ω_2 that yield the diagrams Q_1 and Q_2, respectively. We may then define a map $f : \Omega_1 \to \Omega_2$ by the rule

$$f((A, j)) = (q(A), j \bmod M_2). \quad (3.3.11)$$

Since M_2 divides M_1, f is well-defined, so it remains only to check equivariance. In fact, since f clearly commutes with the action of Z, it suffices to check that f commutes with the action of V_r for $r = 1, 2$. So suppose A and C are seams of Q_1 such that $AV_r = C$, $q(A) = B$, $q(C) = D$, t is the arrow flow between A and C, and u is the arrow flow between B and D (that is, q maps t to u). By the definition of f, we have that

$$f((A, 0)V_r) = f((C, t)) = (D, t \bmod M_2) \quad (3.3.12)$$

and

$$f((A, 0))V_r = (B, 0)V_r = (D, u). \quad (3.3.13)$$

However, from condition (2) of Defn. 3.3.24, we know that $t \equiv u$ (mod M_2). The theorem follows. $\qquad \square$

Among the consequences of the covering theorem is the following result, which is perhaps not entirely obvious from the definitions of quilts and their morphisms.

Corollary 3.3.26. *The composition of two quilt morphisms is a quilt morphism.* $\qquad \square$

Next, we describe the information given by the patches of a quilt, and use that description to define quilts for Seifert groups with $n < \infty$. Recall that in a modular quilt, *patches* (Defn. 3.2.2) correspond precisely with L/R-orbits (Cor. 3.2.17). If we define an L/R-*orbit* of a permutation representation Ω of Σ to be to be a pair (A, AV_2^{-1}), where A is an $\langle L, Z \rangle$-orbit of Ω, and AV_2^{-1} is the associated $\langle R, Z \rangle$-orbit, then the same result still holds.

More precisely, careful analysis shows that the patch analogue of (3.3.1) is actually obtained "for free" from the way we have set things up (Thm. 3.3.28). First, however, we need the following definition.

Definition 3.3.27. Let P be a patch of finite order k in a quilt Q, and let t_{ri} ($r = 1, 2, 1 \leq i \leq k$) be the arrow flows of the dotted 1-cells of P that go into vertices of type r. We define the *inflow* of P to be $-\sum t_{ri}$. Note that the negative sign makes sense with respect to our arrow flow conventions.

It is easy to check that adding boundaries does not change the inflow of a patch, which justifies our saying "the" inflow of a patch of a quilt.

Theorem 3.3.28 (The patch theorem). *Let Ω be a transitive permutation representation of Σ, and let Q be the quilt diagram corresponding to Ω and the Z-orbit representatives $(A, 0)$. Let P be a patch of finite order k in Q, let t_{ri} ($r = 1, 2, 1 \leq i \leq k$) be the arrow flows of the dotted 1-cells of P that go into vertices of type r, and let $t = -\sum t_{ri}$ be the inflow of P.*

Let A (resp. B) be a seam of Q whose left (resp. right) half is in P. Then the smallest positive integer j such that $(A, 0)L^j = (A, \ell)$ (resp. $(B, 0)R^j = (B, \ell)$) for some ℓ is precisely $j = k$. Furthermore,

$$(A, 0)L^k = (A, t), \tag{3.3.14}$$

$$(B, 0)R^k = (B, -t). \tag{3.3.15}$$

Proof. The first claim of the theorem follows from Cor. 3.2.17 and consideration of the modular structure of Q, so it remains to verify (3.3.14) and (3.3.15). We only show (3.3.14), as (3.3.15) is entirely similar.

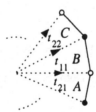

Fig. 3.3.9. Walking one step of L

Now, suppose we have the situation shown in Fig. 3.3.9. Then, taking one step around the patch (that is, applying L once), we get

$$(A, 0)L = (A, 0)V_2^{-1}V_1^{-1} = (B, -t_{21})V_1^{-1} = (C, -t_{21} - t_{11}). \tag{3.3.16}$$

Taking k such steps around P, we get $(A, 0)L^k = \left(A, -\sum t_{ri}\right) = (A, t)$. $\square$

Definition 3.3.29. Let $\Sigma = \left\langle \frac{p_1}{m_1}, \frac{p_2}{m_2}, 0 \right\rangle$ be a Seifert group, let Q be a Σ-quilt, let P be a patch of Q, and let i be the inflow of P. We define the *ramification index* of P to be the additive order of i (mod M). (The ramification index is not to be confused with the collapsing index of a patch of an (m_1, m_2, n)-modular quilt, which is only a matter of $\overline{\Sigma}$-modular structure, not Σ-quilt structure.) If the ramification index of P is 1, we say that P is *unramified*; otherwise, we say that P is *ramified*.

The following is then an immediate consequence of the patch theorem.

Corollary 3.3.30. *Retaining the notation of the patch theorem, suppose P has order k, and suppose that t is the smallest positive integer such that $(A, 0)L^t = (A, 0)$ (resp. $(B, 0)R^t = (B, 0)$). Then t is the product of k and the ramification index of P.* $\square$

In any case, having solidified our understanding of patches, we now extend the definition of quilts to Seifert groups with $n < \infty$. Let $\Sigma = \left\langle \frac{p_1}{m_1}, \frac{p_2}{m_2}, \frac{q}{n} \right\rangle$, $\Sigma_0 = \left\langle \frac{p_1}{m_1}, \frac{p_2}{m_2}, 0 \right\rangle$, and $\overline{\Sigma} = (m_1, m_2, n)$.

Definition 3.3.31. We define a Σ-*quilt* to be a Σ_0-quilt (Q, M) such that:

1. Q is also a $\overline{\Sigma}$-modular quilt; and
2. For every patch of Q with inflow t and collapsing index i (as a $\overline{\Sigma}$-modular quilt), we have

$$ti \equiv q \pmod{M}. \tag{3.3.17}$$

Note that, as we observed after the definition of quilt diagram (Defn. 3.3.4), if (Q, M) is a Σ-quilt, then for every patch of Q of collapsing index i, (3.3.17) and elementary facts about congruences imply that $\gcd(M, i)$ divides q. We therefore have:

Definition 3.3.32. Given a modular quilt Q and a Seifert group Σ, a non-negative integer M is said to be *compatible* with Q and Σ if:

1. For every patch of Q of collapsing index i, $\gcd(M, i)$ divides q; and
2. M satsfies the GCD conditions from Defn. 3.3.6.

If Σ is geometric (Defn. 2.3.3), as before, we have a simpler formulation of compatibility.

Proposition 3.3.33. *Let $\Sigma = \left\langle \frac{p_1}{m_1}, \frac{p_2}{m_2}, \frac{q}{n} \right\rangle$ be a geometric Seifert group, and let Q be a $\overline{\Sigma}$-modular quilt. Then a nonnegative integer M is compatible with Q and Σ if and only if it is relatively prime to the global collapsing index of Q. Furthermore, if M is compatible with Q and Σ, then at any vertex or patch of Q, (3.3.1) and (3.3.17) have unique solutions (mod M).* $\square$

Note that the transitive permutation representations of Σ are precisely those transitive permutation representations of Σ_0 that respect the relation $Z^q = L^n$. Furthermore, it is an easy consequence of the Patch Theorem (Thm. 3.3.28) that a Σ_0-quilt (Q, M) respects the relation $Z^q = L^n$ if and only if

Q is also a $\overline{\Sigma}$-modular quilt and (Q, M) satisifes the patch flow condition (3.3.17). It follows that the bijection theorem for quilts (Thm. 3.3.20) and the covering theorem for quilts (Thm. 3.3.25) extend to Seifert groups with $n < \infty$. Therefore, in the sequel, we use the general case of the bijection and covering theorems freely.

Finally, to end this section, we prove the following theorem, for use in Chaps. 6 and 8.

Theorem 3.3.34 (The pullback theorem). *Let Σ be a Seifert group and let $\overline{\Sigma} = \Sigma/\langle Z \rangle$. If (Q_1, M_1) is a Σ-quilt, and Q_0 is a $\overline{\Sigma}$-modular quilt that covers Q_1, then there exists a unique Σ-quilt (Q_0, M_1) covering (Q_1, M_1).*

Proof. Let $q : Q_0 \to Q_1$ be a modular quilt morphism, and choose a quilt diagram representative for (Q_1, M_1). From Defn. 3.3.24, we see that it is enough to show that there exists a unique arrow flow mod M_1 on Q_0 that is sent by q to the arrow flows of Q_1. Now, if Q_0 is given such an arrow flow, then each dotted 1-cell d of Q_0 must be labelled with the arrow flow on $q(d)$. It is therefore enough to show that pulling an arrow flow on (Q_1, M_1) back to Q_0 produces a quilt diagram. In fact, comparing Defn. 3.3.4, we see that it is enough to show that the pullback flow satisfies (3.3.1) at every vertex of Q_0. (For $n < \infty$, we must also check patches, but the proof is the same, so we omit it.)

So fix $r = 1, 2$, and suppose q maps a type r vertex v_0 of collapsing index i_0 onto a type r vertex v_1 of collapsing index i_1. For $i = 0, 1$, let t_i be the inflow of vertex v_i. Since each dotted 1-cell of vertex v_1 pulls back to $\dfrac{i_1}{i_0}$ dotted 1-cells of v_0, it follows that $t_0 \equiv \dfrac{i_1}{i_0} t_1 \pmod{M_1}$, or in other words, $t_0 i_0 \equiv t_1 i_1 \pmod{M_1}$. Therefore, since (3.3.1) holds at v_1, it holds at v_0 as well. The theorem follows. $\qquad\square$

3.4 Drawing quilts in practice

The reader may be wondering why we have included midlines in seams. One reason is, even though the results of the previous section give a complete picture of how quilt diagrams represent transitive permutation representations of a Seifert group, these diagrams can be awkward to draw in practice. However, as we will show in this section, we can make quilt diagrams much easier to draw by moving most of the arrow flows to the midlines. As we shall see in Chap. 4, this convention is also convenient for our main application of quilts, namely, studying Norton systems.

We would like to define a meaning for quilt diagrams with arrow flows on midlines as well as on dotted 1-cells. Now, by adding flows like those shown in Figs. 3.4.1 and 3.4.2, we can change a quilt diagram with arrow flows on midlines to one without arrow flows on midlines, and vice versa. The flow in

Fig. 3.4.1. Left boundary flow   **Fig. 3.4.2.** Right boundary flow

Fig. 3.4.1 is called a *left boundary flow* of $+j$, and the flow in Fig. 3.4.2 is called a *right boundary flow* of $+j$. Note that we can define (possibly infinite) sums of left and right boundary flows in the same way we defined sums of boundary flows.

Definition 3.4.1. A *pure quilt diagram* is just a quilt diagram as given by Defns. 3.3.4 and 3.3.31. A *non-pure quilt diagram* is a pure quilt diagram with a (well-defined) sum of left and right boundary flows added to it.

Of course, we must check that a non-pure quilt diagram still gives a well-defined quilt.

Proposition 3.4.2. *Two pure quilt diagrams whose flows differ by a sum of left and right boundary flows are homologous.*

Proof. By linearity, it suffices to check this for a single seam; in fact, it is enough to show that if a left boundary flow and a right boundary flow for a given seam cancel on the midline, then their sum is a boundary. However, the sum of a left boundary flow of $+j$ and a right boundary flow of $+k$ produces a flow of $j - k$ on the midline (taking positive flow to be away from dots of type 1 and towards dots of type 2). Therefore, since the sum of a left boundary flow of $+j$ and a right boundary flow of $+j$ is a boundary flow in the previous sense (Defn. 3.3.17), the proposition follows. □

Note that the flow rules (3.3.1) (and (3.3.17), for $n < \infty$) are unaffected by left and right boundaries. It follows that a modular quilt with arrow flows on dotted lines and midlines that satisfies (3.3.1) (and (3.3.17), for $n < \infty$) defines a unique quilt, and it also follows that the results of the previous section extend to non-pure quilt diagrams and quilts. Therefore, from now on, we usually will not distinguish between pure and non-pure quilt diagrams.

The main point of having left and right boundary flows at our disposal is that we can use them to push all of the inflow in a finite patch to a single location. Similarly, we can remove *all* inflow from an infinite patch, a fact we will use later. We also set the convention that we do not draw any dotted 1-cell with no inflow on it. With this convention, moving inflow can considerably improve the appearance of a quilt, so we usually display quilt diagrams in this cleaned-up form. We will also see later that this simplified form still allows us to see all of the key information in a quilt.

Exercise 3.4.3. Consider, once again, the quilt diagram Q shown in Fig. 3.3.3, p. 41. By adding left and right boundaries to Q, show that Q is homologous to the quilt diagram shown in Fig. 3.4.3. (Note that in Fig. 3.4.3, we have omitted all dotted 1-cells with no inflow on them, as mentioned above.)

Fig. 3.4.3. Fig. 3.3.3, cleaned up

3.5 Remarks and references

The reader who is familiar with the modular group $\mathbf{PSL}_2(\mathbf{Z}) \cong (3, 2, \infty)$ may appreciate the following interpretation of $(3, 2, \infty)$-modular quilts. Consider the *quilt tiling*, that is, the tiling of $\mathbf{H}^2$ shown in Fig. 3.5.1. Note that we may use the shaded region of Fig. 3.5.1 as a fundamental region for the (left) action of $\mathbf{PSL}_2(\mathbf{Z})$ on $\mathbf{H}^2$, and that $V_1 = \begin{pmatrix} 1 & 1 \\ -1 & 0 \end{pmatrix}$ and $V_2 = \begin{pmatrix} 0 & -1 \\ 1 & 0 \end{pmatrix}$ act as the indicated isometries of $\mathbf{H}^2$. Let Γ be a subgroup of $\mathbf{PSL}_2(\mathbf{Z})$. Since the action of Γ respects the quilt tiling, the quilt tiling descends to a tiling of $\Gamma\backslash\mathbf{H}^2$. Therefore, if Q is the quilt of Γ, the seams of Q correspond bijectively with the right cosets Γa, which in turn correspond bijectively with the tiles of $\Gamma\backslash\mathbf{H}^2$.

Now, for $r = 1, 2$, the V_r dotted 1-cell of the tile of $\Gamma\backslash\mathbf{H}^2$ corresponding to Γa is attached to the V_r^{-1} dotted 1-cell of the tile corresponding to Γb if and only if there exist $g \in \mathbf{PSL}_2(\mathbf{Z})$, $g_a \in \Gamma a$, and $g_b \in \Gamma b$ such that $gg_a = 1$ and $gg_b = V_r$. However, this occurs if and only if $V_r \in a^{-1}\Gamma b$, or in other words, if and only if $\Gamma a V_r = \Gamma b$. It follows that the tiles of $\Gamma\backslash\mathbf{H}^2$ have the same incidence properties as do the seams of Q. In other words, Q is precisely the quotient $\Gamma\backslash\mathbf{H}^2$ (or in other words, the orbifold $\Gamma\backslash\mathbf{H}^2$), with the usual topology, since we delete infinite patch points.

We remark that permutation representations and coset diagrams have long been used to analyze subgroups of $\mathbf{PSL}_2(\mathbf{Z})$ and other triangle groups. In some sense, this practice goes back as far as Riemann's existence theorem (see Völklein [86, Chap. 4]), but in recent years, some of the earlier references include Millington [59] and Atkin and Swinnerton-Dyer [3, §3]. For an overview, see Jones [47, §2]. More recently, there has been much work on the use of such coset diagrams (or equivalently, what Grothendieck calls *dessins d'enfants*) in algebraic geometry and algebraic number theory, especially as a method for investigating the Galois group of $\overline{\mathbf{Q}}$ (the algebraic

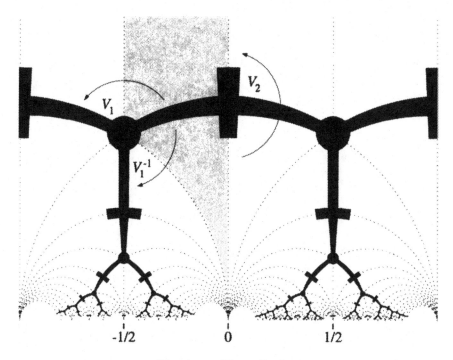

Fig. 3.5.1. The quilt tiling

closure of **Q**) over **Q**. See the survey articles of Jones and Singerman [48] and Schneps [73] for more information; more generally, see the collections edited by Schneps [74, 75].

As for quilts and subgroups of Seifert groups, as mentioned in the introduction, it seems that the closest related previous work is a construction that appears in Conway, Coxeter and Shephard [14]. As we shall see in Sect. 7.1, quilts are a generalization of the dual of one particular case of their construction. For other related work, see Burchenko [10] and Pride and Stohr [71], who use certain annotated complexes to compute Schur multipliers of Coxeter groups, and Baik, Harlander, and Pride [4], who have extended the results of Conway, Coxeter and Shepard [14] from central extensions to arbitrary extensions. It would be quite interesting to pursue a similar non-central variation on quilts; see Sect. 13.7 for some possibilities.

4. Norton systems and their quilts

Quilts (Chap. 3) are especially well-suited to represent the action of $\mathbf{B}_3$ known as *Norton's action* (Sect. 4.2). Specifically, since $\mathbf{B}_3$-quilts correspond bijectively with transitive permutation representations of $\mathbf{B}_3$, we may represent an orbit of Norton's action (a *Norton system*) by its corresponding quilt. More importantly, in practice, it is not hard to draw the quilt of even a fairly sizable Norton system (Sect. 4.3) by formalizing the rules described in Sect. 1.3 of the introduction. We also provide a few examples to illustrate the process of drawing the quilt of a Norton system (Sect. 4.4) and describe the relationship between Norton systems known as *mirror-isomorphism* (Sect. 4.5).

4.1 Summary of Chap. 3 for $\Sigma = \mathbf{B}_3$

Before beginning our study of Norton's action and Norton systems, we first summarize the results of Chap. 3 for the case of the Seifert group $\mathbf{B}_3$. Our notation is from Chaps. 2 and 3. In particular, recall (Sect. 2.4) that

$$\begin{aligned}
\mathbf{B}_3 \cong \langle\, Z, V_1, V_2, L \,|\, 1 = [Z, V_1] = [Z, V_2] = [Z, L], \\
Z = V_1^3, \; Z^{-1} = V_2^2, \\
1 = V_1 V_2 L \,\rangle \\
= \left\langle \frac{1}{3}, -\frac{1}{2}, 0 \right\rangle,
\end{aligned} \qquad (2.4.4)$$

and that by convention, $R = V_2 V_1$. Note that in $\mathbf{B}_3$, $V_1 = R^{-1}L$, and $V_2 = RL^{-1}R$; in fact, $\mathbf{B}_3 \cong \langle L, R \,|\, LR^{-1}L = R^{-1}LR^{-1} \rangle$ (presentation (2.4.2)).

Recall (Defns. 3.2.1 and 3.3.4) that a $\mathbf{B}_3$-*quilt diagram*, or in this chapter, simply a *quilt diagram*, is defined to be a 2-complex Q, a nonnegative integer M, and an *arrow flow* mod M defined on the 1-cells of Q (that is, when Q is finite, a cellular 1-chain mod M), such that:

- Q is made from a collection of *seams* (see Fig. 4.1.1) by identifying each "dotted" V_r 1-cell of a seam with exactly one V_r^{-1} 1-cell ($r = 1, 2$), preserving orientations.
- Q is connected.

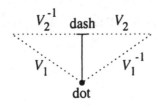

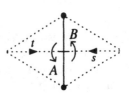

Fig. 4.1.1. A seam of a B_3-quilt

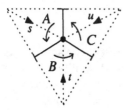

Fig. 4.1.2. Vertex   Fig. 4.1.3. Edge

Fig. 4.1.4. Collapsed vertex   Fig. 4.1.5. Collapsed edge

- The number of seams meeting at a dot (resp. dash) is either 1 or 3 (resp. 1 or 2). The meeting of 3 seams at a dot (resp. 2 seams at a dash) is called a *vertex* (resp. *edge*), as shown in Figs. 4.1.2 and 4.1.3, and a single seam identified with itself around its dot (resp. dash) is called a *collapsed vertex* (resp. *collapsed edge*), as shown in Figs. 4.1.4 and 4.1.5.
- The arrow flow on Q satisfies the following *flow rules*:
  1. The flow into a vertex is 1 mod M. For instance, in Fig. 4.1.2, $s+t+u \equiv 1$ (mod M).
  2. The flow into an edge is -1 mod M; in other words, the flow out of an edge is 1 mod M. For instance, in Fig. 4.1.3, $s+t \equiv -1$ (mod M).
  3. The flow into a collapsed vertex is $\frac{1}{3}$ mod M.
  4. The flow out of a collapsed edge is $\frac{1}{2}$ mod M.

Note that we will sometimes draw flow arrows oriented *out* of edge and collapsed edges, since we may thereby orient all arrows consistently on a non-pure quilt diagram (see Defn. 3.4.1).

Two quilt diagrams are said to be *homologous* if their arrow flows differ by a boundary, and a *quilt* is defined to be a homology class of quilt diagrams (Defn. 3.3.19).

The *quilt representation* induced by a (pure) quilt diagram (Q, M) (Defn. 3.3.10) is a permutation representation of $\mathbf{B}_3$ defined by the following rules.

- The domain of the representation is the set of all pairs (A, j), where A is a seam of Q and j is an integer mod M.
- The action of V_1 (resp. V_2) on the first factor of (A, j) is defined by the circular arrows in Figs. 4.1.2 and 4.1.5 (resp. Figs. 4.1.3 and 4.1.5).
- For $r = 1, 2$, if $AV_r = B$, and the V_r 1-cell between A and B has a flow of t into the attached dash or dot, then

$$(A, j)V_r = (B, j + t). \tag{4.1.1}$$

For example, in Fig. 4.1.3, we have $(A, j)V_2 = (B, j + s)$, and in Fig. 4.1.5, we have $(A, j)V_2 = (A, j + t)$.

Note that if we start with a transitive permutation representation of $\mathbf{B}_3$, after choosing a representative from each Z-orbit, we may reverse the above process and obtain a quilt diagram (Defn. 3.3.15). A different choice of representatives changes the diagram by a boundary, which means that the above operations give a bijective correspondence between quilts and transitive permutation representations of $\mathbf{B}_3$ (Thm. 3.3.20).

Furthermore, the correspondence is natural in the following sense. Let (Q_i, M_i) be quilts $(i = 1, 2)$. If M_2 divides M_1, then a *quilt morphism* (Defns. 3.2.18 and 3.3.24) is a cellular map $q : Q_1 \to Q_2$ that sends vertices to vertices or collapsed vertices, edges to edges or collapsed edges, and so on, and also sends the arrow flow of Q_1 to its reduction mod M_2 on Q_2, for some quilt diagram representatives of Q_1 and Q_2. There exists a morphism from Q_1 to Q_2 if and only if the transitive permutation representation corresponding to Q_1 covers the transitive permutation representation corresponding to Q_2 (Thm. 3.3.25).

Finally, recall that the *patches* of a quilt Q correspond bijectively with the $\langle L, Z \rangle$-orbits of the corresponding transitive permutation representation. More precisely, if A is a seam whose left side is contained in a patch P of order k (Defn. 3.2.5) and *inflow* t (Defn. 3.3.27), then $(A, j)L^k = (A, j + t)$ (Thm. 3.3.28). In particular, if i is the *ramification index* of P (Defn. 3.2.29), then the stabilizer of (A, j) in $\langle L \rangle$ is precisely $\langle L^{ki} \rangle$.

4.2 Norton systems

We begin by defining the action of $\mathbf{B}_3$ known as *Norton's action*. Let G be a group.

Definition 4.2.1. We define an action of $\mathbf{B}_3$ on the set $G \times G$ by the following rules:

$$(\alpha, \beta)L^i = (\alpha, \alpha^i \beta),$$
$$(\alpha, \beta)R^i = (\alpha \beta^i, \beta),$$

(4.2.1)

for $i \in \mathbf{Z}$. We call this action *Norton's action*.

For some motivation behind Defn. 4.2.1, see the introduction (Sect. 1.2), Chap. 10, Mason [56], and Tuite [85]. In any case, we must check the validity of Defn. 4.2.1.

Proposition 4.2.2. *The rules in Defn. 4.2.1 define an action of* $\mathbf{B}_3$.

Proof. Computing, we get

$$(\alpha, \beta)LR^{-1}L = (\alpha\beta^{-1}\alpha^{-1}, \alpha) = (\alpha, \beta)R^{-1}LR^{-1},$$

(4.2.2)

which means that Defn. 4.2.1 respects the defining relation of presentation (2.4.2). □

Definition 4.2.3. An orbit N of the action in Prop. 4.2.2 is called a *Norton system*. If N is finite, the *order* of N is defined to be its size as an orbit. We use $N(\alpha, \beta)$ to denote the Norton system of (α, β).

Note that since the pairs $(\alpha, \beta)R^i = (\alpha\beta^i, \beta)$ and $(\alpha, \beta)L^i = (\alpha, \alpha^i\beta)$ all generate the group $\langle \alpha, \beta \rangle$, it follows that every pair in $N(\alpha, \beta)$ generates $\langle \alpha, \beta \rangle$. We may therefore make the following definition.

Definition 4.2.4. We say that N is a *Norton system for* G if $N = N(\alpha, \beta)$ and $G = \langle \alpha, \beta \rangle$.

Since every Norton system defines a transitive permutation representation of $\mathbf{B}_3$, we define *morphisms*, or *covers*, of Norton systems to be $\mathbf{B}_3$-equivariant maps between them (Defn. 3.1.2). In particular, two Norton systems (possibly for different groups) are *isomorphic* if they are isomorphic as permutation representations of $\mathbf{B}_3$.

One example of a morphism between Norton systems is found in the following theorem.

Theorem 4.2.5. *Let* $G_1 = \langle \alpha, \beta \rangle$ *and* G_2 *be groups, let* φ *be a surjective homomorphism from* G_1 *to* G_2, *and let* $N_1 = N(\alpha, \beta)$ *be a Norton system for* G_1. *Let* f *be the map sending* (α, β) *to* $(\varphi(\alpha), \varphi(\beta))$. *Then* $N_2 = f(N_1)$ *is a Norton system for* G_2, *and* f *is a morphism from* N_1 *to* N_2.

Proof. Since φ is surjective, $\varphi(\alpha)$ and $\varphi(\beta)$ generate G_2. It therefore suffices to show that f is $\mathbf{B}_3$-equivariant, which we need only check for the generators L and R. We have

$$
\begin{aligned}
f((\alpha, \beta)L) &= f((\alpha, \alpha\beta)) \\
&= (\varphi(\alpha), \varphi(\alpha\beta)) \\
&= (\varphi(\alpha), \varphi(\alpha)\varphi(\beta)) \\
&= f((\alpha, \beta))L,
\end{aligned}
$$

(4.2.3)

and a similar result holds for R. The theorem follows.   □

It follows that a group isomorphism φ induces an isomorphism of Norton systems. In particular, conjugate Norton systems for the same group are isomorphic.

Now, since Norton systems are transitive permutation representations of $\mathbf{B}_3$, they may be represented by their corresponding quilts. The covering theorems of Chap. 3 then immediately imply the following corollaries.

Corollary 4.2.6. *Let N_1 and N_2 be Norton systems. N_1 covers N_2 if and only if the stabilizer of a pair in N_1 is contained in the stabilizer of some pair in N_2. In particular, N_1 and N_2 are isomorphic if and only if they have the same conjugacy class of stabilizers in $\mathbf{B}_3$.*   □

Corollary 4.2.7 (Quilts classify Norton systems). *Let N_1 and N_2 be Norton systems. Then N_1 covers N_2 if and only if the quilt of N_1 covers the quilt of N_2. In particular, N_1 and N_2 are isomorphic if and only if their quilts are isomorphic.*   □

Definition 4.2.8. Let G be a group. We say that a quilt Q is a *quilt for G* if Q is the quilt of a Norton system for G (Defn. 4.2.4). Note the distinction between a quilt *for* a group G and the quilt *of* a subgroup or transitive permutation representation of a Seifert group (Defn. 3.3.22).

Of course, the fact that quilts classify Norton systems is less useful if drawing the quilt of a Norton system is more difficult than trying to understand its structure by, say, computer enumeration. Therefore, drawing quilts is the focus of much of the rest of this chapter.

In the rest of this section, we examine the role of $\langle Z \rangle$ in Norton systems. Abstractly, we define the *modulus* of a Norton system to be its modulus as a transitive permutation representation of $\mathbf{B}_3$ (Defn. 3.3.3). However, both to understand and also to calculate the action of $\langle Z \rangle$, it is useful to consider the concept of *stacking*. Fix a Norton system N for a group G. We have the following theorem.

Theorem 4.2.9. *There exists $\kappa \in G$ such that for any $(\alpha, \beta) \in N$, we have*

$$\kappa = \beta^{-1}\alpha\beta\alpha^{-1}.$$

Proof. We verify that κ is the same for the pairs (α, β), $(\alpha, \alpha^i\beta)$, and $(\alpha\beta^i, \beta)$ and any $i \in \mathbf{Z}$, and is therefore invariant under Norton's action.   □

We call κ the *stacking element* for N.

Definition 4.2.10. Given a stacking element $\kappa \in G$, for any $\alpha \in G$ and any integer t, we define α_t by the rules

$$\alpha_0 = \alpha, \tag{4.2.4}$$

$$\alpha_{t+1}\alpha_t\kappa = 1, \tag{4.2.5}$$

where t is any integer.

Example 4.2.11. Let κ be the stacking element of $N(\alpha, \beta)$. In that case, we calculate a few specific examples of Defn. 4.2.10:

$$\alpha_1 = \kappa^{-1}\alpha^{-1} = \left(\alpha^{-1}\right)^{\beta\alpha^{-1}}, \qquad \alpha_{-1} = \alpha^{-1}\kappa^{-1} = \left(\alpha^{-1}\right)^{\beta},$$
$$\alpha_2 = \kappa^{-1}\alpha_1^{-1} = \kappa^{-1}\alpha\kappa = \alpha^{\kappa}, \qquad \alpha_{-2} = \kappa\alpha\kappa^{-1} = \alpha^{\kappa^{-1}}. \tag{4.2.6}$$

Continuing, we see that $\alpha_{t+2} = \alpha_t^{\kappa}$, and that α_n is always conjugate either to α, when n is even, or to α^{-1}, when n is odd. Note that consequently, all of the α_n have the same order.

Next, since $V_1 = R^{-1}L$ and $V_2 = RL^{-1}R$, straightforward computation and an easy induction argument result in the following proposition. (The details are left as an exercise.)

Proposition 4.2.12. *For (α, β) in a Norton system N, we have:*

$$(\alpha, \beta)V_1 = (\alpha\beta^{-1}, \alpha) = ((\alpha^{-1}\beta)_1, \alpha), \tag{4.2.7}$$
$$(\alpha, \beta)V_2 = (\beta, \alpha_{-1}), \tag{4.2.8}$$
$$(\alpha, \beta)Z^i = (\alpha_i, \beta_i), \tag{4.2.9}$$

for any $i \in \mathbf{Z}$. □

Note that (4.2.9) and Exmp. 4.2.11 imply that the modulus of N always divides twice the order of κ. In practice, if G has limited commutativity (for instance, if G is center-free), then the modulus of N will often be equal to either the order of κ or twice the order of κ. Equation (4.2.9) also implies that we may compute the left and right components of $(\alpha, \beta)Z^i$ separately, motivating the following definition.

Definition 4.2.13. Let N be a Norton system, and let $S(N)$ be the set of all elements of G that are part of a pair in N. We define an action of $\langle Z \rangle$ on $S(N)$ by the formula

$$\alpha Z^t = \alpha_t. \tag{4.2.10}$$

(Note that this formula gives a $\langle Z \rangle$-action because of (4.2.9).) An orbit of $S(N)$ under the $\langle Z \rangle$-action is called a *stack*. We say that the elements α_t are *stacked together*, and we call the size of the Z-orbit of α the *stack size* of α, or simply size(α).

Stack sizes in N are closely related to the modulus of N. Specifically, for any (α, β) in N, if size(α) and size(β) are finite, then the modulus of N is equal to the least common multiple of size(α) and size(β), and if either size(α) or size(β) is infinite, then the modulus of N is 0. This is often how the modulus of a Norton system is computed in practice.

4.3 Drawing the quilt of a Norton system

At this point, we have shown that if we take a Norton system $N(\alpha, \beta)$ and consider it as a permutation representation of $\mathbf{B}_3$, then the corresponding quilt describes the structure of $N(\alpha, \beta)$ precisely. In this section, we describe how to draw the quilt of $N(\alpha, \beta)$ from relatively easy computations in $G = \langle \alpha, \beta \rangle$, formalizing the rules from Sect. 1.3 of the introduction. The key to this process is the labelling system we describe here.

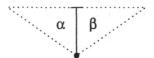

Fig. 4.3.1. A labelled seam

Let N be a Norton system N with modulus M. From Defn. 3.3.15 (see also Sect. 4.1), we see that choosing a representative from each Z-orbit defines a pure quilt diagram Q such that, if (α, β) and (γ, δ) are the representatives for the seams A and B, respectively, $AV_r = B$ ($r = 1, 2$), and the inflow on the dotted 1-cell of type r between A and B is t, then $(\alpha, \beta)V_r = (\gamma, \delta)Z^t$ (see Defn. 3.3.15 or (4.1.1)). In terms of actually drawing Q, we may indicate our choice of representatives by labelling a seam with its representative (α, β) in the manner shown in Fig. 4.3.1. This labelling of the 2-cells of the quilt diagram Q with elements of G is called the *canonical labelling* of Q.

Continuing to apply the results of Chap. 3, we see that the boundary rule (Prop. 3.3.18) means that the effect of replacing (α, β) with (α_i, β_i) is to add a boundary flow of i around the seam of (α, β). Using Defn. 4.2.13, we can extend this boundary rule to left and right boundary flows (Figs. 3.4.1 and 3.4.2, p. 51) by saying that the effect of replacing α (resp. β) with α_i (resp. β_i) is to add a left (resp. right) boundary flow of $+i$ to the seam of (α, β). For example, this extended boundary rule implies that adding a left boundary of $+t$ to Fig. 4.3.1 gives Fig. 4.3.2.

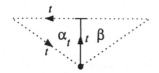

Fig. 4.3.2. Fig. 4.3.1, plus a boundary

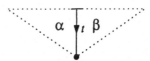

Fig. 4.3.3. A seam (α, t, β)

Using these rules, we define a canonically labelled *non-pure* quilt diagram to be a quilt diagram that differs from a canonically labelled pure quilt diagram by adding a boundary. We refer to the seams of a canonically labelled (non-pure) quilt diagram by using the triple (α, t, β) to denote a seam whose left label is α, whose right label is β, and whose midline flow is t, oriented as shown in Fig. 4.3.3. Note that the notation (α, t, β) says nothing about the flows on the dotted 1-cells of the seam.

We now interpret the features of a quilt in terms of certain relations in G, with the goal of describing how relatively simple computations in G can be used to draw a quilt. We begin with the relations at a seam.

Proposition 4.3.1. *Let Q be a canonically labelled quilt diagram of a Norton system N. If Q contains a seam (α, t, β), then*

$$(\alpha_t, \beta) \in N. \tag{4.3.1}$$

Proof. It suffices to verify that the proposition holds for pure quilt diagrams, and then check that the validity of the proposition is invariant under addition of a boundary. In the pure case, $t = 0$, so the proposition clearly holds. Adding a left boundary of $+j$ changes α to α_j, t to $t - j$, and leaves β unchanged, so the validity of the proposition is indeed invariant under addition of a left boundary, since $((\alpha_j)_{t-j}, \beta) = (\alpha_t, \beta)$. A similar result holds for adding a right boundary, and the proposition follows. □

The method of verifying an assertion for pure quilt diagrams and then checking that the validity of the proposition is invariant under addition of a boundary serves to prove several results in this section. Therefore, from now on, we will just say that such results "follow from the standard argument" and omit the details.

Next, we consider edges.

Proposition 4.3.2. *Let Q be a canonically labelled quilt diagram of a Norton system N. If the V_2 dotted line of a seam (α, t, β') touches the V_2^{-1} dotted line of a seam (β, s, α'), and the dotted line where they join is labelled with an outflow of v, as shown in Figs. 4.3.4 and 4.3.5, then*

$$(\alpha_t, \beta')V_2 = (\beta_s, \alpha')Z^{-v-s}. \tag{4.3.2}$$

In particular,

$$\beta'_v = \beta. \tag{4.3.3}$$

Proof. In the pure case ($s = t = 0$), by the definition of the quilt representation, $(\alpha, \beta')V_2 = (\beta, \alpha')Z^{-v}$, so (4.3.2) follows from the standard argument. As for (4.3.3), (4.2.8) implies that $(\alpha_t, \beta')V_2 = (\beta', \alpha_{t-1})$, so (4.2.9) and (4.3.2) imply that $\beta' = \beta_{-v}$. □

For the uncollapsed edge shown in Fig. 4.3.4, Prop. 4.3.2 means that

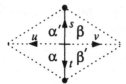

Fig. 4.3.4. Edge   **Fig. 4.3.5.** Collapsed edge

$$(\alpha_t, \beta')V_2 = (\beta_s, \alpha')Z^{-v-s}, \tag{4.3.4}$$

$$(\beta_s, \alpha')V_2 = (\alpha_t, \beta')Z^{-u-t}, \tag{4.3.5}$$

$$\alpha'_u = \alpha, \tag{4.3.6}$$

$$\beta'_v = \beta. \tag{4.3.7}$$

Similarly, for the collapsed edge in Fig. 4.3.5,

$$(\alpha_t, \alpha')V_2 = (\alpha_t, \alpha)Z^{-\frac{1}{2}}, \tag{4.3.8}$$

$$\alpha'_v = \alpha. \tag{4.3.9}$$

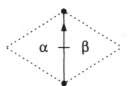

Fig. 4.3.6. Basic edge

Note that, for any $(\alpha, \beta) \in N$ whose seam is in the uncollapsed edge in Fig. 4.3.4, by adding boundaries, we can set $t = u = v = 0$. Since the flow rules then imply that s must equal 1, we see that by adding boundaries, we can make the canonical labelling of any given uncollapsed edge look like the *basic edge* in Fig. 4.3.6. Basic edges are the formal version of what we called "directed edges" in the introduction (Sect. 1.3). More precisely:

Corollary 4.3.3 (The basic edge principle). *Unless $(\alpha, \beta)V_2 = (\alpha_i, \beta_i)$ for some i mod M, we may assume that the quilt of $N(\alpha, \beta)$ contains the basic edge shown in Fig. 4.3.6.* □

Note that it is often easy to check that $(\alpha, \beta)V_2$ does not equal (α_i, β_i) for any i mod M. For instance, this holds if α and β have different orders, or more generally, if α is conjugate to neither β nor β^{-1}.

Proceeding to vertices, we have the following proposition.

Proposition 4.3.4. *Let Q be a canonically labelled quilt diagram of a Norton system N. If the V_1 dotted line of a seam (β, r, α') touches the V_1^{-1} dotted*

*line of a seam (γ, s, β'), and the dotted line where they join is labelled with
an inflow of v, as shown in Figs. 4.3.7 and 4.3.8, then*

$$(\beta_r, \alpha')V_1 = (\gamma_s, \beta')Z^{v+r}. \tag{4.3.10}$$

In particular,

$$\beta'_v = \beta. \tag{4.3.11}$$

Proof. In the pure case ($r = s = 0$), from the definition of the quilt representation, $(\beta, \alpha')V_1 = (\gamma, \beta')Z^v$, so (4.3.10) follows from the standard argument. Then, since (4.2.7) implies $(\beta_r, \alpha')V_1 = (\beta_r\alpha'^{-1}, \beta_r)$, (4.3.10) implies that $\beta'_v = \beta$.   □

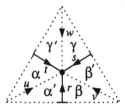

Fig. 4.3.7. Vertex

Fig. 4.3.8. Collapsed vertex

For the uncollapsed vertex in Fig. 4.3.7, Prop. 4.3.4 says that

$$(\alpha_t, \gamma')V_1 = (\beta_r, \alpha')Z^{u+t}, \tag{4.3.12}$$
$$(\beta_r, \alpha')V_1 = (\gamma_s, \beta')Z^{v+r}, \tag{4.3.13}$$
$$(\gamma_s, \beta')V_1 = (\alpha_t, \gamma')Z^{w+s}, \tag{4.3.14}$$
$$\alpha'_u = \alpha, \tag{4.3.15}$$
$$\beta'_v = \beta, \tag{4.3.16}$$
$$\gamma'_w = \gamma. \tag{4.3.17}$$

Similarly, for Fig. 4.3.8,

$$(\alpha_t, \alpha')V_1 = (\alpha_t, \alpha')Z^{\frac{1}{3}}, \tag{4.3.18}$$
$$\alpha'_v = \alpha. \tag{4.3.19}$$

Next, we have the following simple but important corollary to Prop. 4.3.4.

Corollary 4.3.5 (The multiplication principle). *Let Q be a canonically labelled quilt for N, with (in the notation of Fig. 4.3.7) a vertex such that $r \equiv 1 \pmod{M}$ and $s \equiv t \equiv u \equiv v \equiv w \equiv 0 \pmod{M}$. Then $\alpha' = \alpha$, $\beta' = \beta$, and $\gamma' = \gamma = \alpha\beta$.*

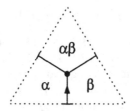

Fig. 4.3.9. Multiplication vertex

Fig. 4.3.9 illustrates a vertex of the type described in the statement of Cor. 4.3.5. Such a vertex is called a *multiplication vertex*.

Proof. Everything is immediate from (4.3.15)–(4.3.17), except for $\gamma = \alpha\beta$. However, (4.3.12) implies that $(\alpha, \gamma)V_1 = (\beta_1, \alpha)$, so

$$
\begin{aligned}
(\alpha, \gamma) &= (\beta_1, \alpha)V_1^{-1} \\
&= (\alpha\beta^{-1}\alpha^{-1}, \alpha)L^{-1}R \\
&= (\alpha\beta^{-1}\alpha^{-1}, \alpha\beta)R \\
&= (\alpha, \alpha\beta). \quad \square
\end{aligned}
\tag{4.3.20}
$$

Since every vertex is homologous to a multiplication vertex, the multiplication principle means that every vertex can be interpreted as a multiplication relation among the elements of $S(N)$. Consequently, the multiplication principle is one of the keys to working with quilts of Norton systems.

Having interpreted seams, edges, and vertices of the quilt of a Norton system for G in terms of relations in G, we now consider the converse problem, namely, how to use relations in G to determine information about the quilt of a Norton system for G. The following is the main result.

Theorem 4.3.6. *Let N be a Norton sytem with stacking element κ, modulus M, and quilt $Q(N)$, and let (Q, M) be a quilt diagram whose 2-cells are labelled with elements of G. If the labels of (Q, M) satisfy the seam (4.3.1), edge (4.3.2), and vertex (4.3.10) relations, then Q covers $Q(N)$. If, in addition, Q has just one seam corresponding to every Z-orbit of N, then Q is a canonically labelled quilt diagram in the class of $Q(N)$.*

Proof. First, note that, since we know the stacking element κ, α_i is well-defined for all $\alpha \in G$. Furthermore, since the labels of Q satisfy the seam relations, for every seam of Q labelled (α, t, β), (α_t, β) is in N, and size(α) and size(β) divide M. Therefore, without loss of generality, we may add boundaries to Q as we wish, changing the labels accordingly. In particular, we may assume that Q is a pure quilt diagram.

Let $\Omega = \{(A, j)\}$ be the representation induced by the quilt diagram Q, and let f be the map defined by

$$f((A,i)) = (\alpha_i, \beta_i), \qquad\qquad (4.3.21)$$

where $(\alpha, 0, \beta)$ is the label on A. Because the labels on Q respect the seam relations, f is a well-defined map from Ω to N. Therefore, to obtain the first statement of the theorem, it is enough to show that f is $\mathbf{B}_3$-equivariant, or in other words, that $f((A,i)V_r) = f((A,i))V_r$ for any $(A,i) \in \Omega$, $r = 1, 2$.

So, for $(A,i) \in \Omega$ and $r = 1, 2$, let $(B,j) = (A,i)V_r$, and suppose that A and B are labelled $(\alpha, 0, \beta)$ and $(\gamma, 0, \delta)$, respectively. From the definition of the quilt representation, if t is the flow into the edge or vertex at which A and B meet, then $j = i + t$. Therefore,

$$\begin{aligned}
f((A,i)V_r) &= f((B,j)) \\
&= (\gamma_j, \delta_j) \\
&= (\gamma_i, \delta_i)Z^t \qquad\qquad (4.3.22)\\
&= (\alpha_i, \beta_i)V_r \\
&= f((A,i))V_r,
\end{aligned}$$

where the fourth equality follows from either the edge (4.3.2) or the vertex (4.3.10) relations. The first statement of the theorem follows.

As for the last statement of the theorem, if Q has just one seam corresponding to every Z-orbit of N, then Q must cover $Q(N)$ with degree 1 (Defn. 3.1.4), since Q and $Q(N)$ have the same modulus. Therefore, Q is isomorphic to $Q(N)$. $\qquad\square$

In short, to draw the quilt of a Norton system $N(\alpha, \beta)$, we assemble a diagram Q that works "locally", and then check to make sure that the Z-orbits represented by any two seams of Q are distinct.

Our final tool for making and understanding quilts of Norton systems is the patch theorem (Thm. 3.3.28). Now, since (4.3.3) and (4.3.11) imply that the labels in a patch are all stacked together (Defn. 4.2.13), each patch has a unique associated stack. We may therefore define a *stack element* for a patch P to be an element of the stack associated with P. The following corollary then describes how the order of a patch relates to the order of its stack elements.

Corollary 4.3.7 (Patch theorem). *Let (Q, M) be the quilt of a Norton system N, let α be a stack element for a patch P of Q, and let k and i be the order and inflow of P, respectively. Then we have two cases:*

1. *If k is finite, the order of α is k times the ramification index of P.*
2. *If k is infinite, the order of α is infinite.*

We also have that $\text{size}(\alpha)$ divides i. In particular, if $\text{size}(\alpha) = M$, then P is unramified.

Proof. For any such α, pick some β such that (α, β) is contained in the L-orbit associated with P. Now, since $(\alpha, \beta)L^j = (\alpha, \alpha^j\beta)$, the order of the action of L on (α, β) is precisely the order of α. The corollary then follows from Thm. 3.3.28 (see the end of Sect. 4.1 for a summary), except for the last statement. However, since (3.3.14) implies that

$$(\alpha, \alpha^k \beta) = (\alpha, \beta)L^k = (\alpha_i, \beta_i), \tag{4.3.23}$$

we see that $\alpha = \alpha_i$ and size(α) divides i.   □

Remark 4.3.8. As mentioned in Sect. 3.4, by adding boundaries, we can move all of the inflow of a patch to one of its dotted 1-cells. It then follows from (4.3.3) and (4.3.11) that we can always label every patch of the quilt of a Norton system with just one of its stack elements. Since this practice greatly improves the appearance of a quilt, we will always do so.

4.4 Examples and exercises

Example 4.4.1. Consider the quilt Q of $N = N(1,0)$ for $G = C_5$, which we write as the integers mod 5. Since G is abelian, $\kappa = 0$, and for $\alpha \in S(N)$, $\alpha_i = (-1)^i \alpha$. Furthermore, since $1 \neq -1$ in C_5, $M = 2$.

We start, naturally, with the seam of $(1,0)$. Since 0 is not stacked with 1, we may assume that Q contains the basic edge shown in Fig. 4.4.1.

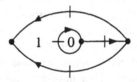

Fig. 4.4.1. First step of Example 4.4.1   Fig. 4.4.2. Second step of Example 4.4.1

Since $M = 2$, the only possible patch ramification indices are 1 and 2, so the patch theorem (Cor. 4.3.7) implies that the patch of 1 must have order 5 and the patch of 0 must have order 1, with both patches unramified. Using the fact that either 1 or 3 seams meet at a vertex (a rule we call, by a slight abuse of terminology, the *trivalent rule*), and applying the flow rules, without loss of generality, we come to the situation shown in Fig. 4.4.2. Note that in Fig. 4.4.2, we are forced to identify the ends of the edge in Fig. 4.4.1.

Since the vertex at the right hand side of Fig. 4.4.2 is a multiplication vertex, the "outside" patch (the one including the point at infinity) must be labelled 2. This patch must also have order 5, so another application of the trivalent rule and the flow rules gets us to the situation shown in Fig. 4.4.3.

One last application of the multiplication principle allows us to fill in the remaining patch with 0, and we arrive at our final answer (Fig. 4.4.4). We

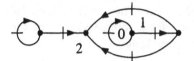

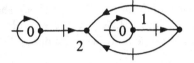

Fig. 4.4.3. Third portion of
Example 4.4.1

Fig. 4.4.4. Final answer for
Example 4.4.1

leave it as an exercise for the reader to check that the seams of Fig. 4.4.4 do
not represent any Z-orbits of N more than once.

Exercise 4.4.2. Using the methods of Example 4.4.1, draw the quilts $N(1,0)$
for the cyclic groups of order greater than 5 and less than 12. (C_1 through
C_4 are a little trickier, since all of those quilts have some kind of collapse.)
C_{11} is a particularly interesting case, since its quilt has genus 1. The answers
can be found in Sect. 5.3.

Example 4.4.3. Let Q be the quilt of $N((0\ 1\ 2),(0\ 1\ 2\ 3\ 4))$, a Norton system for $G = A_5$. A calculation shows that $\kappa = (0\ 2\ 3)$ and $M = 3$.

Once again, our basic method for drawing Q is to use the trivalent rule,
the flow rules, the multiplication principle, and the patch theorem. However,
this time we will just assume that all patches are unramified unless a problem
occurs. Now, on the one hand, patches with stack size M are never ramified,
so this assumption is often likely to be true. On the other hand, since any
result we get using the unramified patch assumption will at least *cover* Q, we
need only check, once we have a completed diagram, that all the seams of our
diagram are distinct. In fact, it is often sufficient to check that all (or even
most) of the patches are associated with distinct stacks. This is therefore our
basic method for making many quilts, especially quilts for perfect or almost
perfect groups, which tend to have fewer collapses.

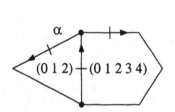

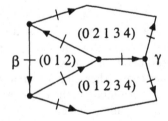

Fig. 4.4.5. Beginning of Example
4.4.3

Fig. 4.4.6. Second step of Example
4.4.3

In any case, given the unramified patch assumption, we begin with the
situation shown in Fig. 4.4.5. Now, the vertex involving $(0\ 1\ 2)$, $(0\ 1\ 2\ 3\ 4)$,
and α is a multiplication vertex, so

$$\alpha = (0\ 1\ 2)(0\ 1\ 2\ 3\ 4) = (0\ 2\ 1\ 3\ 4). \tag{4.4.1}$$

If we again assume no collapse, we arrive at Fig. 4.4.6.

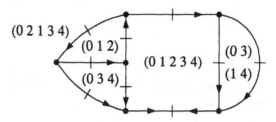

Fig. 4.4.7. Answer for Example 4.4.3

Proceeding with β and γ as we did for α, we get

$$\beta = (0\ 1\ 2)(0\ 2\ 1\ 3\ 4) = (0\ 3\ 4), \tag{4.4.2}$$

$$\gamma = (0\ 2\ 1\ 3\ 4)(0\ 1\ 2\ 3\ 4) = (0\ 3)(1\ 4). \tag{4.4.3}$$

Filling in the rest of the diagram using the trivalent rule, the multiplication principle, and the flow rules, and rearranging things slightly, we get the result shown in Fig. 4.4.7.

We leave it to the reader to check that the seams of Fig. 4.4.7 are distinct, which implies that Fig. 4.4.7 actually is the quilt of N. The reader may wish to compare the process used in the introduction (Sect. 1.3).

The idea we have used to draw Exmps. 4.4.1 and 4.4.3 ("make patches until the quilt closes up") can actually be turned into a theorem, as we shall see in Sect. 8.2. The key to making this idea work turns out to be that both of these quilts have genus 0, and both have no collapse. Quilts with lots of collapse and quilts of higher genus can be much harder to draw.

Exercise 4.4.4. Draw the quilt of the Norton system of $((0\ 1\ 2), (0\ 1))$ for the group S_3. This is harder than it sounds, and should provide a good test for the reader's understanding of the material presented so far. For the answer, see Fig. 5.3.7.

4.5 Mirror-isomorphism

To end this chapter, we prove some results used elsewhere (Chap. 10 and [44]). Recall that there is an outer automorphism Φ of $\mathbf{B}_3$ (Sect. 2.4, after presentation (2.4.2)) that exchanges L and R, and let π^Φ denote the image of $\pi \in \mathbf{B}_3$ under Φ. We have the following theorems.

Theorem 4.5.1. (α, β) and $(\alpha^{-1}, \beta^{-1})$ have the same stabilizer in $\mathbf{B}_3$, and therefore, have isomorphic Norton systems.

Proof. Let $\Delta = \text{Stab}(\alpha, \beta)$. We have that

$$\text{Stab}((\alpha, \beta)Z) = Z^{-1}\Delta Z = \Delta. \tag{4.5.1}$$

However,

$$\begin{aligned}
(\alpha, \beta)Z &= (\alpha\beta^{-1}\alpha^{-1}\beta\alpha^{-1}, \alpha\beta^{-1}\alpha^{-1}) \\
&= ((\beta\alpha^{-1})^{-1}\alpha^{-1}(\beta\alpha^{-1}), (\beta\alpha^{-1})^{-1}\beta^{-1}(\beta\alpha^{-1})).
\end{aligned} \tag{4.5.2}$$

Since the structure of a Norton system is invariant under conjugation (Thm. 4.2.5), the theorem follows. $\square$

We say that the Norton systems N_1 and N_2 are *mirror-isomorphic* if, for some injective $f : N_1 \to N_2$, $f((\alpha, \beta)\pi) = f((\alpha, \beta))\pi^\Phi$. Note that the quilts of mirror-isomorphic Norton systems are "isomorphic" by an orientation-reversing mapping.

Theorem 4.5.2. *The Norton system of (α, β) is mirror-isomorphic to the Norton system of (β, α).*

Proof. Let τ be the map that switches the first and second entries of a pair of elements of G, and let -1 be the map that inverts each entry of a pair of elements of G. The reader may easily check that the following diagram commutes.

$$
\begin{array}{ccccc}
(\alpha, \beta) & \overset{\tau}{\longleftrightarrow} & (\beta, \alpha) & \overset{-1}{\longleftrightarrow} & (\beta^{-1}, \alpha^{-1}) \\
\Big\downarrow {\scriptstyle L} & & & & \Big\downarrow {\scriptstyle R} \\
(\alpha, \alpha\beta) & \overset{\tau}{\longleftrightarrow} & (\alpha\beta, \alpha) & \overset{-1}{\longleftrightarrow} & (\beta^{-1}\alpha^{-1}, \alpha^{-1})
\end{array}
\tag{4.5.3}
$$

From (4.5.3), and from the analogous diagram obtained by switching the roles of L and R, we see that the Norton system of (α, β) is mirror-isomorphic to that of $(\beta^{-1}, \alpha^{-1})$. The theorem then follows from Thm. 4.5.1. $\square$

Thm. 4.5.2 implies that mirror-image quilts are essentially the same. Therefore, in the sequel, when classifying a set of quilts or Norton systems, we often do so up to mirror-isomorphism.

5. Examples of quilts

We present some examples of quilts, concentrating mainly on the quilts of Norton systems. Our main goals are to show how to make and draw such quilts, and to establish examples for later use. Some other examples are discussed in [44], and some larger examples are the focus of Chap. 10.

All quilts in this chapter are $\mathbf{B}_3$-quilts, and most are quilts for some finite group G. All modular quilts in this chapter are $\mathbf{PSL}_2(\mathbf{Z})$-quilts.

5.1 Naming and drawing quilts

Since we need a way to name quilts, we take this opportunity to describe our standard notation. The symbol

$$M \cdot S^g(a_t^i, b_u^j, c_v^k, \dots) \tag{5.1.1}$$

denotes a quilt of modulus M, S seams, genus g, and i patches of order a and inflow t, j patches of order b and inflow u, k patches of order c and inflow v, and so on.

By convention, if $g = 0$, it is omitted in (5.1.1). Inflows of 0 are omitted, and a^1 is abbreviated as a. In text, we sometimes further abbreviate (5.1.1) to $M \cdot S$, leaving the genus, patch orders, and inflow unspecified.

Fig. 5.1.1. Example of identification notation

Notation 5.1.1. We sometimes indicate the surface on which a quilt is drawn by identifying the midpoints (dashes) of various quilt edges. For example, in Fig. 5.1.1, the two midpoints marked (a) are identified in an orientation-preserving manner, as are the two midpoints marked (b). In general, once these identifications have been made, it is easy to compute the genus of the surface by counting patches, edges, and vertices. For example, the quilt in Fig. 5.1.1 has exactly 1 patch (why?), and therefore has genus 1.

5.2 The B_3-quilts with fewer than 6 seams

In this section, we enumerate the B_3-quilts with 5 or fewer seams, both to provide some examples of small quilts and to describe how we may enumerate quilts in general. The quilts in this section are not necessarily quilts of Norton systems; see Sect. 9.1 for a description of which are and which are not.

Let s be a positive integer. To enumerate the quilts with s seams, we first enumerate the *modular* quilts with s seams. There are several methods for doing this, either by hand or by machine; see, for instance, Atkin and Swinnerton-Dyer [3, §3.3]. One such method is to list all of the ways the s seams may be partitioned into collapsed and uncollapsed vertices, and then determine all possible ways these collections of collapsed and uncollapsed vertices may be attached inside edges or collapsed edges.

Example 5.2.1. Suppose that Q is a modular quilt Q with 4 seams. Now, the seams of Q must either be in 4 collapsed vertices or 1 vertex and 1 collapsed vertex. Clearly, it is impossible to join 4 collapsed vertices in edges to make a connected surface, so only the latter case can occur. In that case, because of connectedness, the collapsed vertex must be joined with the uncollapsed vertex in an edge, so the only choice to be made is whether the remaining unattached seams are both in a collapsed edge, or joined to each other in an uncollapsed edge. It follows that there are two possible modular quilts with 4 seams, namely, those shown in Figs. 5.2.5 and 5.2.6, below.

Once we have enumerated the modular quilts with s seams, for each modular quilt Q, we then must determine all possible arrow flows mod M on Q, up to homology, for all nonnegative integers M that are compatible with Q. By adding boundaries, we can assume that all of the inflow of a patch is located at a single edge or collapsed edge. Using this assumption, and applying the flow rules (see Defn. 3.3.4 or Sect. 4.1), we can often deduce what many (though in general, not all) of the arrow flows on midlines must be. More precisely, as we will see later (Cor. 6.1.12), if Q has genus g and p patches, then the homology classes of arrow flows on Q are determined by $2g + p - 1$ freely chosen parameters mod M.

Example 5.2.2. Suppose that Q is the (unique) modular quilt with 2 seams (Fig. 5.2.2). Since Q has collapsed vertices but no collapsed edges, M is compatible with Q if and only if $\gcd(3, M) = 1$. Now, by adding boundaries, we can assume that the flow into the lone patch of Q occurs at a single location, as shown in Fig. 5.2.2. However, from the flow rules, the flow into each of the collapsed vertices must be $1/3$ (mod M). Therefore, since the flow out of the lone edge of Q must be 1 (mod M), the flow into the patch of Q must be $1/3$ (mod M). It follows that the quilt in Fig. 5.2.2 is the unique quilt with 2 seams and modulus M.

By applying the above techniques, we see that the quilts with 5 or fewer seams are precisely those shown in Figs. 5.2.1–5.2.7. The variable j, when it appears, denotes a freely chosen parameter mod M.

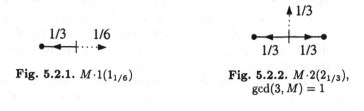

Fig. 5.2.1. $M \cdot 1(1_{1/6})$

Fig. 5.2.2. $M \cdot 2(2_{1/3})$,
gcd$(3, M) = 1$

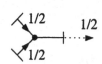

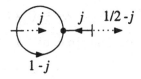

Fig. 5.2.3. $M \cdot 3(3_{1/2})$,
gcd$(2, M) = 1$

Fig. 5.2.4. $M \cdot 3(1_j, 2_{1/2-j})$,
gcd$(2, M) = 1$

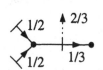

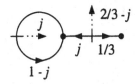

Fig. 5.2.5. $M \cdot 4(4_{2/3})$,
gcd$(6, M) = 1$

Fig. 5.2.6. $M \cdot 4(1_j, 3_{2/3-j})$,
gcd$(3, M) = 1$

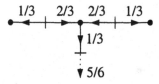

Fig. 5.2.7. $M \cdot 5(5_{5/6})$, gcd$(6, M) = 1$

5.3 The groups of order < 12

The quilts for the groups of order < 12 are shown in Figs. 5.3.1–5.3.17, except for the quilt for C_5, which we have already seen in Sect. 4.4. All quilts are canonically labelled (Sect. 4.3). As it happens, the groups of order < 12 have just one isomorphism class of quilts apiece. In addition, note that C_{11} is the smallest group with a quilt of genus greater than 0. The genus 1 quilt for C_{11} can be obtained from the region in Fig. 5.3.17 by identifying its boundary in an orientation-preserving manner as indicated.

In this section, we use the following conventions for naming group elements. An element of C_n is denoted by an integer mod n, and an element of $C_m \times C_n$ is denoted by an ordered pair of an integer mod m and an integer mod n. The dihedral group of order $2n$ is written as

$$D_{2n} = \langle t, x \,|\, 1 = x^n = t^2,\ t^{-1}xt = x^{-1} \rangle. \tag{5.3.1}$$

The quaternion group of order 8 is written as

$$Q_8 = \langle i, j, k \,|\, i^2 = j^2 = k^2 = ijk \rangle. \tag{5.3.2}$$

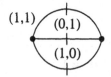

Fig. 5.3.1. C_1: $1 \cdot 1(1)$

Fig. 5.3.2. C_2: $1 \cdot 3(1, 2)$

Fig. 5.3.3. C_3: $2 \cdot 4(1, 3)$

Fig. 5.3.4. C_4: $2 \cdot 6(1, 1_2, 4)$

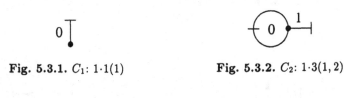

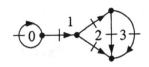

Fig. 5.3.5. $C_2 \times C_2$: $1 \cdot 6(2^3)$

Fig. 5.3.6. C_6: $2 \cdot 12(1, 2, 3, 6)$

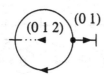

Fig. 5.3.7. S_3: $3 \cdot 3(1_{-1}, 2)$

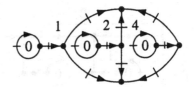

Fig. 5.3.8. C_7: $2 \cdot 24(1^3, 7^3)$

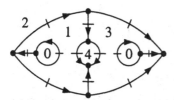

Fig. 5.3.9. C_8: $2 \cdot 24(1^2, 2, 4, 8^2)$

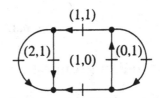

Fig. 5.3.10. $C_4 \times C_2$: $2 \cdot 12(2^2, 4^2)$

Fig. 5.3.11. D_8: $2 \cdot 6(2^2, 2_1)$

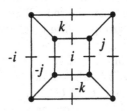

Fig. 5.3.12. Q_8: $1 \cdot 24(4^6)$

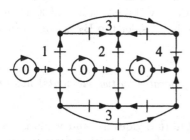

Fig. 5.3.13. C_9: $2 \cdot 36(1^3, 3^2, 9^3)$

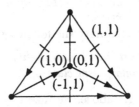

Fig. 5.3.14. $C_3 \times C_3$: $2 \cdot 12(3^4)$

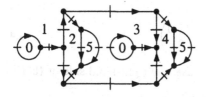

Fig. 5.3.15. C_{10}:
$2 \cdot 36(1^2, 2^2, 5^2, 10^2)$

Fig. 5.3.16. D_{10}: $5 \cdot 3(1_{-2}, 2)$

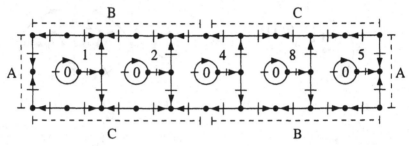

Fig. 5.3.17. C_{11}: $2 \cdot 60^1 (1^5, 11^5)$

5.4 Abelian groups

The Norton systems for an abelian group have an especially simple structure. In this section, we show that all Norton systems for an abelian group are isomorphic (Thm. 5.4.1) and that the collapsing in a quilt for an abelian group is very limited (Thm. 5.4.2). We also describe how quilts for finite abelian groups relate to *congruence subgroups* of $\mathbf{SL}_2(\mathbf{Z})$ (Thm. 5.4.3).

Throughout this section, we write all abelian groups additively.

Recall that an abelian group generated by two elements is isomorphic to $\mathbf{Z}/a \times \mathbf{Z}/b$, with b dividing a. (Either a, or both a and b, may be 0.) If we write elements of $G \cong \mathbf{Z}/a \times \mathbf{Z}/b$ as column vectors with the top entry in $\mathbf{Z}/a$ and the bottom entry in $\mathbf{Z}/b$, pairs of elements of G can be considered as 2×2 matrices with the entries of the top row in $\mathbf{Z}/a$ and the entries of the bottom row in $\mathbf{Z}/b$. In this context, the right action of $\pi \in \mathbf{B}_3$ on a pair of elements of G is just right multiplication by the projection of π down to $\mathbf{SL}_2(\mathbf{Z})$. In other words, we may effectively consider the $\mathbf{B}_3$-action to be an $\mathbf{SL}_2(\mathbf{Z})$-action.

We then have the following theorem.

Theorem 5.4.1. *An abelian group G (generated by two elements) has only one isomorphism class of Norton system.*

Proof. As above, we take $G \cong \mathbf{Z}/a \times \mathbf{Z}/b$ (b divides a), and write pairs of elements of G as 2×2 matrices. It is enough to show that for any pair of elements $\begin{pmatrix} x & y \\ w & z \end{pmatrix}$ that generates G, there exists $\pi \in \mathbf{SL}_2(\mathbf{Z})$ such that

$$\begin{pmatrix} x & y \\ w & z \end{pmatrix} \pi = \begin{pmatrix} 1 & 0 \\ 0 & d \end{pmatrix} \tag{5.4.1}$$

with d relatively prime to b, for in that case, the right-hand side of (5.4.1) is equivalent to $\begin{pmatrix} 1 & 0 \\ 0 & 1 \end{pmatrix}$ by an automorphism of $\mathbf{Z}/b$.

Now, since reduction mod a is a surjective homomorphism from $\mathbf{SL}_2(\mathbf{Z})$ to $\mathbf{SL}_2(\mathbf{Z}/a)$ (Schoeneberg [76, IV §2,1]), we can treat the right action of

$\mathbf{SL}_2(\mathbf{Z})$ as if it were an action by $\mathbf{SL}_2(\mathbf{Z}/a)$. Since $\gcd(x, y, a) = 1$, we may choose m, n (mod a) such that $mx + ny \equiv 1$ (mod a). In that case,

$$\begin{pmatrix} x & y \\ w & z \end{pmatrix} \begin{pmatrix} m & -y \\ n & x \end{pmatrix} = \begin{pmatrix} 1 & 0 \\ c & d \end{pmatrix} \tag{5.4.2}$$

for some $c, d \in \mathbf{Z}/b$. Now, a linear combination of the columns of the right-hand side of (5.4.2) that is equal to 0 must involve a trivial multiple of the first column, mod a. Therefore, $\begin{pmatrix} 1 & 0 \\ c & d \end{pmatrix}$ generates a group $\mathbf{Z}/a \times \mathbf{Z}/n$, where n is the additive order of d mod b. However, since $n = b$, d must be relatively prime to b, so multiplying (5.4.2) by $\begin{pmatrix} 1 & 0 \\ -ce & 1 \end{pmatrix}$, where $de \equiv 1$ (mod b), achieves the desired result. $\square$

We can therefore talk about "the" quilt for an abelian group. By abuse of terminology, we call such a quilt an *abelian quilt*. Note that if G is cyclic, the above argument actually shows that G has just one Norton system.

One feature of abelian quilts is that they are almost always without collapse. Let (Q, M) be the quilt of a Norton system N for an abelian group G. First, since the action of $Z^2 \in \mathbf{B}_3$ is conjugation, M divides 2. Therefore, if Q has a collapsed edge, we must have $M = 1$, and there is some pair (α, β) in the associated Norton system such that

$$(\alpha, \beta)E = (\alpha, \beta). \tag{5.4.3}$$

In other words, $-\beta = \alpha$ and $\alpha = \beta$, which means that G is a homomorphic image of C_2. Similarly, if Q has a collapsed vertex, there is some pair (α, β) such that

$$(\alpha - \beta, \alpha) = (\alpha, \beta)V = (\alpha, \beta)_1 = (-\alpha, -\beta), \tag{5.4.4}$$

which implies that $\alpha = \beta$ and $3\alpha = 0$. Therefore, G is a homomorphic image of C_3.

The other cases of collapse are similar. Specifically, we leave it as an exercise for the reader to show that if Q has a ramified patch, then G is a homomorphic image of C_4, and that if Q has modulus 1, then G is a homomorphic image of $C_2 \times C_2$. We may therefore sum up our discussion of collapse in abelian quilts with the following theorem.

Theorem 5.4.2. *If Q is a quilt for an abelian group of order ≥ 5, then Q has no collapsed edges or vertices, and no unramified patches. Furthermore, the modulus of Q is 2.* $\square$

We end this section by explaining the relationship between abelian quilts and congruence subgroups of the modular group $\mathbf{SL}_2(\mathbf{Z})$.

Theorem 5.4.3. *A quilt for a finite abelian group represents a conjugacy class of congruence subgroups of $\mathbf{SL}_2(\mathbf{Z})$.*

Proof. Let Q be the quilt for $\mathbf{Z}/a \times \mathbf{Z}/b$ (preserving the notation from before Thm. 5.4.1). Since the modulus of Q divides 2, Q represents a conjugacy class of subgroups of $\mathbf{SL}_2(\mathbf{Z})$. Also, we can assume that $a = b$, since the quilt for $\mathbf{Z}/a \times \mathbf{Z}/a$ covers the quilt for $\mathbf{Z}/a \times \mathbf{Z}/b$, and a group containing a congruence subgroup is congruence. Therefore, it remains to show that the quilt for $\mathbf{Z}/N \times \mathbf{Z}/N$ represents a conjugacy class of congruence subgroups. Equivalently, we can examine the stabilizer of the pair $\begin{pmatrix} 1 & 0 \\ 0 & 1 \end{pmatrix}$ (with all entries in $\mathbf{Z}/N$). That stabilizer is precisely those elements of $\mathbf{SL}_2(\mathbf{Z})$ equal to $\begin{pmatrix} 1 & 0 \\ 0 & 1 \end{pmatrix}$ mod N, or in other words, $\Gamma(N)$, the principal congruence subgroup of level N (Schoeneberg [76, IV §2,1]). □

In particular, the quilt for $C_N \times C_N$ is the quilt of $\Gamma(N)$. (For $N = 3$, 4, or 5, this quilt is the tetrahedron, cube, or dodecahedron, respectively.) Similarly, the quilt for C_N is the quilt of $\Gamma_1(N)$. On the other hand, there are many congruence subgroups that do not correspond to abelian quilts. For instance, when $\Gamma_0(N) \neq \Gamma_1(N)$, $\Gamma_0(N)$ does not correspond with an abelian quilt, since an abelian quilt with an unramified patch of order 1 must arise from a cyclic group C_N, and the quilt for C_N is the quilt of $\Gamma_1(N)$.

5.5 Exercises

Exercise 5.5.1. Draw the quilts for the groups of order 12. As it happens, each of these groups has just one isomorphism class of quilts. The answers are described in Table 5.5.1. (Note that $3{:}4$ denotes the semidirect product $\langle \alpha, \beta \, | \, 1 = \alpha^3 = \beta^4, \beta^{-1}\alpha\beta = \alpha^{-1} \rangle$.)

Table 5.5.1. Quilts for the groups of order 12

Group	Quilt
C_{12}	$2{\cdot}48(1^2, 2, 3^2, 4^2, 6, 12^2)$
$C_2 \times C_6$	$2{\cdot}24(2^3, 6^3)$
D_{12}	$6{\cdot}6(2^2, 2_1)$
A_4	$4{\cdot}4(1_2, 3)$
$3{:}4$	$6{\cdot}6(1_{-1}, 1_2, 4)$

Exercise 5.5.2. Draw the quilts for S_6. It can be shown (by, for example, computer enumeration) that S_6 has 4 isomorphism classes of quilts. Table 5.5.2 gives a sample generating pair and brief description for each class.

Table 5.5.2. Quilts for S_6

Quilt	Pair
$5 \cdot 30(2, 4^4, 6^2)$	$(0\ 1\ 2)(3\ 4), (2\ 3)(4\ 5)$
$6 \cdot 36(2, 3, 4^2, 5, 6^3)$	$(0\ 1\ 2\ 3\ 4), (4\ 5)$
$8 \cdot 48(2, 4^3, 5^2, 6^4)$	$(0\ 1\ 2\ 3\ 4\ 5), (0\ 2\ 4\ 3\ 1\ 5)$
$10 \cdot 60(3^2, 4^2, 5^2, 6^6)$	$(0\ 1\ 2\ 3\ 4\ 5), (2\ 3\ 5\ 4)$

Exercise 5.5.3. Draw the quilts for A_6. Table 5.5.3 gives a sample generating pair and brief description for each of the 4 isomorphism classes. These are a little harder to draw than the quilts of the previous exercise, since all but the $4 \cdot 24$ quilt have some kind of collapse. Pictures of the $5 \cdot 15$ and $4 \cdot 24$ quilts can be found in Sect. 11.5.

Table 5.5.3. Quilts for A_6

Quilt	Pair
$10 \cdot 10(1_5, 4, 5)$	$(0\ 1\ 2\ 3\ 4), (2\ 3)(4\ 5)$
$5 \cdot 15(3^2, 4, 5)$	$(0\ 1\ 2)(3\ 4\ 5), (2\ 4\ 5)$
$8 \cdot 16(2, 4, 5^2)$	$(0\ 1)(2\ 3\ 4\ 5), (1\ 3)(4\ 5)$
$4 \cdot 24(3^2, 4^2, 5^2)$	$(0\ 1\ 2\ 3)(4\ 5), (0\ 2)(1\ 4\ 5\ 3)$

Exercise 5.5.4. Let Q be the $4 \cdot 24$ quilt for A_6. Show that if a group G has a quilt isomorphic to Q, then there exist $\alpha, \beta \in G$ such that $G = \langle \alpha, \beta \rangle$ and

$$1 = \alpha^4 = \beta^4 = (\alpha\beta)^5 = (\alpha^{-1}\beta)^5 = (\alpha^2\beta)^3 = (\alpha\beta^2)^3. \tag{5.5.1}$$

Hint: can you "see" these relations in the patches of the quilt Q?

Exercise 5.5.4 is actually part of a very general phenomenon. In fact, as we shall see in Sects. 8.2 and 8.3, the relations (5.5.1) in the group G are enough to completely determine the structure of the quilt Q for G.

6. The combinatorics of quilts

In this chapter, we present some combinatorial results on quilts. First, in Sect. 6.1, by applying elementary algebraic topology, we classify all quilts with a given modular stucture, obtaining some other useful lifting theorems in the process. Then, in Sect. 6.2, by applying elementary combinatorial geometry, we obtain counting formulas for modular quilts.

6.1 Lifting conditions for quilts

Throughout this section, we fix a Seifert group $\Sigma = \left\langle \frac{p_1}{m_1}, \frac{p_2}{m_2}, \frac{q}{n} \right\rangle$ and its projection $\overline{\Sigma} = (m_1, m_2, n) \bmod \langle Z \rangle$.

Consider the following problem.

Problem 6.1.1. Given a $\overline{\Sigma}$-modular quilt Q and a nonnegative integer M, which arrow flows mod M on Q give a Σ-quilt diagram $\widetilde{Q}$? If there exists a valid choice of arrow flows, how many different quilts $\widetilde{Q}$ are there with modular structure Q, and how can they be described?

Stated in terms of subgroups, Prob. 6.1.1 becomes:

Problem 6.1.2. Given a subgroup $\Gamma \le \overline{\Sigma}$ and a nonnegative integer M, is there a subgroup $\Delta \le \Sigma$ such that Δ projects down to Γ and $\Delta \cap \langle Z \rangle = \langle Z^M \rangle$? If so, how many such Δ are there, and how can they be described?

A fortiori, an answer to Prob. 6.1.2 also solves the following problem.

Problem 6.1.3. Given a subgroup $\Gamma \le \overline{\Sigma}$, classify all subgroups $\Delta \le \Sigma$ that project down to Γ.

As we will see, the answer to the first part of Probs. 6.1.1–6.1.2 comes from 0-homology (Thms. 6.1.6 and 6.1.7), and the answer to the second part of Probs. 6.1.1–6.1.2 comes from 1-homology (Thms. 6.1.10 and 6.1.17). We begin with the following concepts.

Definition 6.1.4. Let X be a finite graph, and let M be a nonnegative integer. An *arrow flow* mod M on X is defined to be a 1-chain on X with values in $\mathbf{Z}/M$, and an *inflow specification* mod M on X is defined to be a

0-chain on X with values in $\mathbf{Z}/M$. (See Sect. 2.5 for a brief review of the homology of 2-complexes.) The *total inflow* $\tau(S)$ of an inflow specification S is defined to be the sum of the values of S. Finally, we say that an inflow specification S has a *flow solution* mod M if there exists some arrow flow a such that $S = \partial a$, where ∂ is the usual boundary homomorphism.

The most interesting case of Defn. 6.1.4 is when X is the 1-skeleton of a finite modular quilt Q. In that case, if a is an arrow flow on X, ∂a assigns to each (dot or patch point) 0-cell of X the inflow at that 0-cell resulting from a. Furthermore, suppose S is the inflow specification given by the flow rules of a Σ-quilt with modulus M. From Defn. 3.3.4, we see that if t is the value of S at a vertex of type r ($r = 1, 2$) and collapsing index i, then

$$ti \equiv p_r \pmod{M}. \tag{6.1.1}$$

If we also have $n < \infty$, then it follows from the patch theorem (Thm. 3.3.28), that if t is the value of S at a patch of collapsing index i, then

$$ti \equiv q \pmod{M}. \tag{6.1.2}$$

Therefore, if X is the 1-skeleton of a modular quilt, we define a *quilt inflow specification* on X to be an inflow specification on X that satisfies (6.1.1) at vertices, and if $n < \infty$, satisfies (6.1.2) at patches.

Recall also that if Σ is geometric (Defn. 2.3.3), and M is compatible with Q (Defns. 3.3.6 and 3.3.32), then p_r/i is well-defined mod M, and condition (6.1.1) becomes $t \equiv p_r/i \pmod{M}$. Similarly, if Σ is geometric and $n < \infty$, then (6.1.2) becomes $t \equiv q/i \pmod{M}$. Therefore, if Σ is geometric with $n = \infty$, then a $\overline{\Sigma}$-modular quilt Q and a modulus M compatible with Q determine a unique quilt inflow specification, except at the patches of Q; and if Σ is geometric with $n < \infty$, then a $\overline{\Sigma}$-modular quilt Q and a modulus M compatible with Q determine a quilt inflow specification completely.

In any case, having established our framework, we now apply some basic results on 0-homology (see Sect. 2.5).

Proposition 6.1.5. *Let X be a finite connected graph. An inflow specification S on X has a flow solution mod M if and only if $\tau(S) \equiv 0 \pmod{M}$.*

Proof. Since every 0-chain is a 0-cycle, the sequence

$$C_1(X; \mathbf{Z}/M) \xrightarrow{\partial} C_0(X; \mathbf{Z}/M) \longrightarrow H_0(X; \mathbf{Z}/M) \tag{6.1.3}$$

is exact, and since X is connected, Thm. 2.5.8 implies that the second arrow in (6.1.3) is precisely τ, the sum of the values of a 0-chain. However, since the inflow specifications with flow solutions are precisely those in the image of ∂, which is precisely the kernel of τ, the theorem follows.   □

Interpreting Prop. 6.1.5 in terms of the 1-skeleton of a quilt, we obtain a necessary and sufficient condition (Thm. 6.1.6) for a quilt inflow specification on a finite modular quilt to have an arrow flow solution mod M. Recall (Defn. 2.3.2) that the *Euler number* $e(\Sigma) \in \mathbf{Q}$ of Σ is defined to be

$$
e(\Sigma) = \begin{cases} \dfrac{p_1}{m_1} + \dfrac{p_2}{m_2} + \dfrac{q}{n} & \text{if } n < \infty, \\[2mm] \dfrac{p_1}{m_1} + \dfrac{p_2}{m_2} & \text{if } n = \infty. \end{cases} \tag{6.1.4}
$$

The following theorem exhibits one way in which $e(\Sigma)$ measures the extent to which Σ is different from the direct product $\overline{\Sigma} \times \langle Z \rangle$. We first consider the case where $n = \infty$.

Theorem 6.1.6 (Lifting theorem). *Let Σ be a Seifert group with $n = \infty$. Let Q be a finite $\overline{\Sigma}$-modular quilt, let M be a nonnegative integer, let s be the number of seams in Q, and let I be the global collapsing index of Q (Defn. 3.2.8). Then $sIe(\Sigma)$ is an integer, and for any $u \in \mathbf{Z}/M$, the following are equivalent:*

1. *There exists a quilt $\widetilde{Q}$ with modular structure Q, modulus M, and total patch inflow u;*
2. *M is compatible with Q and $\overline{\Sigma}$, and*

$$
uI + sIe(\Sigma) \equiv 0 \pmod{M}. \tag{6.1.5}
$$

Note that since most of the details of the proof arise from the presence of collapse in Q, the reader who wants to concentrate on the main ideas may prefer to consider the case $I = 1$ at first.

Proof. For $r = 1, 2$, let $\mathbf{v}_r$ be the set of all (possibly collapsed) vertices of type r in Q, and for v a vertex of either type, let $i(v)$ be the collapsing index of v.

We begin by counting seams and vertices. Now, for $r = 1, 2$, every seam is contained in a unique vertex of type r, and a type r vertex v of collapsing index $i(v)$ contains $m_r/i(v)$ seams. Therefore, by summing over $\mathbf{v}_r$, we get

$$
\sum_{v \in \mathbf{v}_r} \frac{m_r}{i(v)} = s. \tag{6.1.6}
$$

Multiplying both sides of this equation by I/m_r, we obtain

$$
\frac{sI}{m_r} = \sum_{v \in \mathbf{v}_r} \frac{I}{i(v)} \tag{6.1.7}
$$

Note that since I is the least common multiple of all collapsing indices of Q, both sides of (6.1.7) are integers. It follows that

$$
sIe(\Sigma) = \frac{sIp_1}{m_1} + \frac{sIp_2}{m_2} \tag{6.1.8}
$$

is also an integer.

Next, we show that if S is a quilt inflow specification mod M for Q, then the left-hand side of (6.1.5) is equal to $I\tau(S)$ (mod M). Fix $r = 1, 2$. Now, for every vertex v of Q of type r, let $t(v)$ be the inflow specified by S at v. Summing $t(v)$ over $\mathbf{v}_r$ and multiplying by I, we have

$$I \sum_{v \in \mathbf{v}_r} t(v) = \sum_{v \in \mathbf{v}_r} \frac{I}{i(v)} t(v) i(v) \qquad (6.1.9)$$

where again, $I/i(v)$ is always an integer. However, since S is a quilt inflow specification, and (6.1.1) implies that $t(v)i(v) \equiv p_r$ (mod M), we see that

$$I \sum_{v \in \mathbf{v}_r} t(v) \equiv \sum_{v \in \mathbf{v}_r} \frac{I}{i(v)} p_r \quad (\text{mod } M). \qquad (6.1.10)$$

Applying (6.1.7), we get

$$I \sum_{v \in \mathbf{v}_r} t(v) \equiv p_r \frac{sI}{m_r} \quad (\text{mod } M). \qquad (6.1.11)$$

Then, since the total inflow of S is the inflow at vertices plus the total patch inflow u, using (6.1.11) for $r = 1, 2$, we get

$$\begin{aligned} I\tau(S) &= I \left(u + \sum_{r=1,2} \sum_{v \in \mathbf{v}_r} t(v) \right) \\ &\equiv uI + \frac{sIp_1}{m_2} + \frac{sIp_1}{m_2} \quad (\text{mod } M) \\ &\equiv uI + sIe(\Sigma) \quad (\text{mod } M). \end{aligned} \qquad (6.1.12)$$

Turning to the equivalence of conditions (1) and (2) of the theorem, suppose that condition (1) holds. Certainly M is compatible with Q and $\overline{\Sigma}$, so it remains to verify (6.1.5). However, if S is the inflow specification of a quilt $\tilde{Q}$ with modular structure Q, Prop. 6.1.5 implies that $\tau(S) \equiv 0$ (mod M), so certainly $I\tau(S) \equiv 0$ (mod M). Equation (6.1.5) then follows from (6.1.12).

Conversely, given condition (2) of the theorem, since M is compatible with Q and $\overline{\Sigma}$, there exists some quilt inflow specification S mod M for Q. From (6.1.12), we know that $I\tau(S) \equiv 0$ (mod M), so by elementary facts on congruences, if $M_0 = \dfrac{M}{\gcd(I, M)}$, then $\tau(S) \equiv 0$ (mod M_0). In other words, for some integer k,

$$\tau(S) \equiv kM_0 \quad (\text{mod } M). \qquad (6.1.13)$$

If $\gcd(I, M) = 1$, then $M_0 = M$, and we are done; however, if $\gcd(I, M) > 1$, we may have to modify our choice of S. Again by elementary facts on congruences, examining (6.1.1), we see that at any vertex of collapsing index i, we may add an arbitrary multiple of $\dfrac{M}{\gcd(i, M)}$ to the inflow specified by S and still get a valid quilt inflow specification. However, since I is by definition the least common multiple of all collapsing indices i, by the Chinese Remainder Theorem, we may find another quilt inflow specification S' such that $\tau(S')$ is $\tau(S)$ plus an arbitrary multiple of M_0. In particular, we may cancel out the original "error" of kM_0, and the theorem follows.   □

Exactly the same proof, with the consideration of the "free" inflow u at patches replaced by calculations like those we did at vertices, yields:

Theorem 6.1.7 (Lifting theorem). *Let Σ be a Seifert group with $n < \infty$. Let Q be a finite $\overline{\Sigma}$-modular quilt, let s be the number of seams in Q, and let I be the global collapsing index of Q (Defn. 3.2.8). Then $sIe(\Sigma)$ is an integer, and the following are equivalent:*

1. *There exists a quilt $\widetilde{Q}$ with modular structure Q and modulus M;*
2. *M is compatible with Q and $\overline{\Sigma}$, and*

$$sIe(\Sigma) \equiv 0 \pmod{M}. \quad \square \qquad (6.1.14)$$

Thms. 6.1.6 and 6.1.7 give a complete solution to the first question of Probs. 6.1.1 and 6.1.2. In particular, we obtain the following very useful condition on possible moduli for lifts of a fixed modular quilt.

Corollary 6.1.8. *Let Σ be a Seifert group with $n = \infty$, let Q be a Σ-quilt with $s < \infty$ seams and modulus M, and suppose Q has no ramified patches. Then M divides $sIe(\Sigma)$.*   □

In other words, if we think of a quilt as a sort of bundle over a surface, purely "horizontal" information about Q (number of seams, collapsing index) implies "vertical" information about Q (modulus). This "horizontal implies vertical" principle is the key to much of Chap. 8.

Example 6.1.9. Let $\Sigma = \mathbf{B}_3 = \langle \frac{1}{3}, -\frac{1}{2}, 0 \rangle$ and $\overline{\Sigma} = (3, 2, \infty)$, and consider a finite $\overline{\Sigma}$-modular quilt Q with s seams. If I is the collapsing index of Q, then Thm. 6.1.6 implies that there exists a quilt $\widetilde{Q}$ with modular structure Q and modulus M, and no ramified patches, if and only if M divides $\dfrac{sI}{6}$ (since $e(\Sigma) = -\frac{1}{6}$).

In terms of subgroups, consider a subgroup Γ of finite index s in $\mathbf{PSL}_2(\mathbf{Z})$. If Δ is a subgroup of $\mathbf{B}_3$ that projects down to Γ, we say that Δ *splits over conjugates of* $\langle L \rangle$ if, for $\pi \in \Delta$, $\pi^{-1} L^n \pi$ is an element of Δ if and only if $\pi^{-1} L^n \pi$ projects down to an element of Γ. Thm. 6.1.6 implies that if I is the least common multiple of the orders of all torsion elements of Γ, then there exists a subgroup $\Delta \leq \mathbf{B}_3$ that projects down to Γ, splits over conjugates of $\langle L \rangle$, and satisfies $\Delta \cap \langle Z \rangle = \langle Z^M \rangle$ if and only if M divides $\dfrac{sI}{6}$.

Again, while the condition of having no ramified patches (splitting over conjugates of $\langle L \rangle$) may not seem natural at the moment, the important point is that given this condition, the "horizontal" information of Q (or Γ) yields "vertical" information about $\widetilde{Q}$ and its modulus M.

We now turn to the second question of Probs. 6.1.1 and 6.1.2, which we restate as follows: Let Q be a finite modular quilt, and let S be a quilt inflow specification on Q. If S has some quilt flow solution, say a_0, how can we characterize *all* such solutions?

The key observation is that if a is any other flow solution for S, then $a - a_0$ is a flow solution for the inflow specification 0. Conversely, adding a flow solution for 0 to a_0 results in another flow solution for S. We see that the set of differences between a_0 and any other flow solution for S forms an abelian group, which we denote by $D(S, a_0)$. In fact, since a flow solution for 0 is just a 1-cycle on Q, $D(S, a_0)$ is precisely $Z_1(Q, \mathbf{Z}/M)$, the 1-cycles on Q.

Now, $D(S, a_0)$ is not yet the solution space we want, since solutions that differ by a boundary flow produce the same quilt. Our solutions are therefore characterized by $D(S, a_0)$ modulo boundaries, which immediately leads to the following theorem.

Theorem 6.1.10. *Let Q be a finite modular quilt, let S be a quilt inflow specification on Q that has a flow solution a_0, and let $\bar{a}_0$ be the homology class of a_0. Then the distinct flow solutions for S are precisely the arrow flows of the form $\bar{a}_0 + a$, where a is an element of $H_1(Q; \mathbf{Z}/M)$.*   □

From the universal coefficient theorem, we have that

$$H_1(Q; \mathbf{Z}/M) \cong (\mathbf{Z}/M)^{2g}, \tag{6.1.15}$$

where g is the genus of Q. In particular, when $g = 0$, we have the following corollary.

Corollary 6.1.11 (Uniqueness of genus zero flow solutions). *If Q is a finite modular quilt of genus 0, then a quilt inflow specification on Q has at most one flow solution.*   □

In particular, for a modular quilt Q of genus 0, if Σ is geometric and $n = \infty$, then there exists at most one quilt of modulus M with modular structure Q and specified patch inflows, and if Σ is geometric and $n < \infty$, then there exists at most one quilt of modulus M with modular structure Q.

Let Σ be a geometric Seifert group with $n = \infty$. In that case, since a modular quilt completely determines its quilt inflow specification at vertices, we can combine Thms. 6.1.6 and 6.1.10 in the following statement.

Corollary 6.1.12. *Let Σ be a geometric Seifert group with $n = \infty$, let Γ be a subgroup of finite index in $\overline{\Sigma}$, and let Q be the quilt of Γ. If Q has p patches and genus g, then the lifts of Γ to Σ with modulus M are classified by the following $\mathbf{Z}/M$-valued parameters:*

1. *The patch inflows of Q, which are p parameters satisfying a single linear relation (in other words, p − 1 free parameters); and*
2. *An "affine space" of solutions, specified by 2g free parameters.*   □

If Σ is geometric and $n < \infty$, then a result analogous to Cor. 6.1.12 holds, except without the patch parameters, since patch inflows are determined by modular structure.

Remark 6.1.13. For a more concrete interpretation of the $2g$ $\mathbf{Z}/M$-valued parameters that appear in (6.1.15) and Cor. 6.1.12, see Rem. 8.4.8.

Now, even though Thms. 6.1.6, 6.1.7, and 6.1.10 are essentially results in elementary homology theory, they have many useful consequences. For instance, we will use the following fact in Sect. 12.2.

Theorem 6.1.14. *Let Q be the quilt of a Norton system for an abelian group of order ≥ 5. Then 12 divides the number of seams of Q.*

Proof. From Thm. 5.4.2, we know that Q has no collapsed edges or vertices, or ramified patches, and that Q has modulus 2. Since $d(Q) = 6$, the theorem follows from Cor. 6.1.8.   □

A more important consequence of Thms. 6.1.6, 6.1.7, and 6.1.10 is the genus 0 lifting principle described in Thm. 6.1.17. To state this principle, we first need the following definition.

Definition 6.1.15. For $i = 1, 2$, let (Q_i, M_i) be a quilt. Suppose that M_2 divides M_1 and f is a modular morphism from Q_1 to Q_2. We say that f is a *patch inflow covering* if, for every patch p_1 of order k_1 in Q_1 that covers a patch p_2 of order k_2 in Q_2, we have

$$t(p_1) \equiv \left(\frac{k_1}{k_2} \right) t(p_2) \pmod{M_2}, \tag{6.1.16}$$

where $t(p_r)$ is the inflow at p_r $(r = 1, 2)$.

Exercise 6.1.16. Show that every quilt morphism is a patch inflow covering. Similarly, show that if Σ is a Seifert group with $n < \infty$, then every *modular* quilt morphism from (Q_1, M_1) to (Q_2, M_2) such that M_2 divides M_1 is an inflow covering. (Hint: compare the proof of Thm. 3.3.34.)

Theorem 6.1.17. *Let Σ be a geometric Seifert group, and let (Q_i, M_i) be a Σ-quilt of genus 0, for $i = 1, 2$. Suppose M_2 divides M_1 and f is an inflow covering from Q_1 to Q_2. Then f is actually a quilt morphism.*

Proof. Now, the map sending Q_1 to its reduction mod M_2 is a quilt morphism, and *a fortiori*, f is an inflow covering from the reduction of Q_1 (mod M_2) to Q_2. Therefore, since the composition of two quilt morphisms is a quilt morphism (Cor. 3.3.26), it is enough to consider the case where $M_1 = M_2$.

Let S_1 (resp. S_2) be the inflow specification on Q_1 (resp. Q_2), and let Q_2' be a quilt diagram in the class of Q_2. If the pullback of the arrow flows of Q_2' to Q_1 (Thm. 3.3.34) were a flow solution for S_1, the theorem would follow, because the fact that Q_1 has genus 0 implies that there is a *unique* homology class of flow solutions for S_1.

It therefore suffices to show that if f is an inflow covering from Q_1 to Q_2, where Q_1 and Q_2 have the same modulus M, then the pullback of an arrow flow solution for S_2 is an arrow flow solution for S_1. This is simply a matter of comparing solutions at patches and vertices. Considering patches first, suppose f maps patch p_1 to patch p_2, and suppose that k_1 (resp. k_2) is the order of p_1 (resp. p_2). Every 1-cell of p_2 pulls back to (k_1/k_2) 1-cells of p_1, so the amount of inflow at p_1 in the pullback is (k_1/k_2) times the inflow at p_2. However, from condition (6.1.16), this is precisely what the inflow must be to solve S_1, so the assertion holds for patches.

As for vertices, suppose, for $r = 1, 2$, that f maps a type r vertex v_1 of Q_1 with collapsing index i_1 to a type r vertex v_2 of Q_2 with collapsing index i_2. Since Σ is geometric, the inflow specified at vertex v_1 is precisely p_r/i_1 (mod M), and the inflow at vertex v_2 is precisely p_r/i_2 (mod M). However, since every 1-cell touching v_2 pulls back to i_2/i_1 1-cells touching v_1, the pullback gives a flow solution at v_1. The theorem follows.   □

Thm. 6.1.17 is very useful for determining whether one quilt of genus 0 covers another, since we now only have to check that all of the patch inflows map correctly, instead of having to find explicit quilt diagram representatives for each quilt that map correctly.

Finally, it behooves us to say a few words about lifting conditions for quilts and modular quilts with an infinite number of seams. The two problems that arise are dealing with infinite 0-chains and 1-chains, and dealing with infinite patches. To take care of these problems, we consider the dual complex to the quilt triangulation. Fig. 6.1.1 shows an example of an infinite quilt (including the shaded region), along with its dual complex. Dashed and dotted lines represent the usual solid and dotted 1-cells of the quilt, respectively, and solid lines represent the dual complex. The principal feature of Fig. 6.1.1 is (part of) an infinite patch, represented as the top half of the plane.

Note that the topology of the quilt changes when we go from the quilt triangulation to its dual. For example, the shaded region in Fig. 6.1.1 is part of the quilt, but is not part of its dual complex. Nevertheless, it is easy to see that the new space is a deformation retract of the old one. We leave it to the interested reader to show that the dual complex may be oriented so everything we have phrased in terms of homology becomes phrased in terms of cohomology, changing dimension n to dimension $2 - n$. For a quilt inflow specification S to have a solution, then, it is sufficient for S to be a 2-coboundary. However, since $H^2(Q; \mathbf{Z}/M)$ is 0, as can be shown by (for example) a direct limit argument and the universal coefficient theorem, every

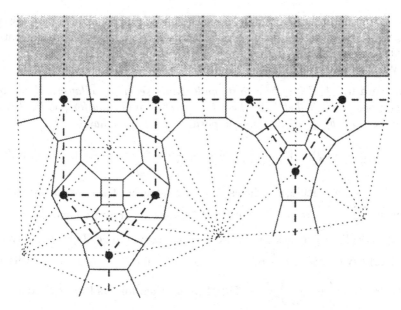

Fig. 6.1.1. The dual complex of a quilt

quilt inflow specification has a flow solution. Similarly, we see that the solution space of an inflow specification is classified by $H^1(Q; \mathbf{Z}/M)$.

6.2 Curvature formulas

In this section, we describe some combinatorial formulas obtained by applying the combinatorial Gauss-Bonnet theorem (see Sect. 2.7) to $\mathbf{PSL}_2(\mathbf{Z})$-modular quilts. (Similar results may be obtained for other kinds of modular quilts, but we will not use them.) This can be thought of as a generalization of the genus zero argument of Norton [62]. Throughout this section, as in Sect. 2.7, we use revs as our units of angle and curvature, where 1 rev = 2π.

Fig. 6.2.1. Angles on a geometric seam

The main idea is that, by giving every seam the geometric structure of a $(3\ 2\ n)$ $(0 < n \leq \infty)$ triangle, we may use the combinatorial Gauss-Bonnet theorem (Thm. 2.7.3) to get counting theorems for modular quilts. In other words, if Q is a (modular) quilt, on each side of every seam of Q, we assign angles of $\frac{1}{4}$ revs at the dash, $\frac{1}{6}$ revs at the dot, and $\frac{1}{2n}$ revs at the patch point, as shown in Fig. 6.2.1.

We then have the following result.

Theorem 6.2.1. *Let Q be a finite quilt with s seams, e collapsed edges, and v collapsed vertices. Let P be the set of patches of Q, and for $p \in P$, denote the order of p by $|p|$. Then for any positive $n \leq \infty$, we have*

$$\chi(Q) = \left(\frac{1}{n} - \frac{1}{6} \right) s + \frac{1}{2} e + \frac{2}{3} v + \sum_{p \in P} \left(1 - \frac{|p|}{n} \right). \qquad (6.2.1)$$

Proof. First, a calculation shows that the integral of the curvature over a seam is $\left(\dfrac{1}{n} - \dfrac{1}{6} \right)$ revs (see Defn. 2.7.2). As for curvature concentrated at 0-cells, uncollapsed edges and vertices have 0 curvature, collapsed eges (resp. vertices) have curvature $\dfrac{1}{2}$ revs (resp. $\dfrac{2}{3}$ revs), and a patch point of order $|p|$ has curvature $\left(1 - \dfrac{|p|}{n} \right)$ revs. The theorem then follows from the combinatorial Gauss-Bonnet theorem (Thm. 2.7.3). $\qquad \square$

As it turns out, (6.2.1) gives essentially the same information no matter what n is. Nevertheless, we state our results by their "dependence" on n, to highlight the different type of geometry involved in each result, and to make the proofs clearer. For instance, the case of $n \leq 5$ gives the following.

Corollary 6.2.2. *Let Q be a quilt with s sides, and suppose that every patch of Q has order at most n. Then for $n = 2, 3, 4,$ or 5, we have that $s \leq 6, 12, 24,$ or 60, respectively, and Q has genus 0.*

Proof. We apply (6.2.1) with the above n, and with $|p| \leq n$ for all patches p. All of the terms on the right side of (6.2.1) are nonnegative, and $\left(\dfrac{1}{n} - \dfrac{1}{6} \right)$ is positive. Therefore, $\chi(Q)$ must be 2, and s must be less than or equal to 2 divided by $\left(\dfrac{1}{n} - \dfrac{1}{6} \right)$, which leads to the corollary. $\qquad \square$

Note that a maximal s in Cor. 6.2.2 can only occur when there are no collapsed edges or vertices and every patch has order n. In fact, the maxima of Cor. 6.2.2 are realized precisely by the appropriate regular polyhedra.

The most interesting case of Thm. 6.2.1 is when $n = 6$. In that case, the curvature of a quilt is concentrated at its 0-cells, leading to the following corollary.

Corollary 6.2.3 (Counting formula for $n = 6$). *Let Q be a quilt such that every patch of Q has at most 6 sides. Then Q has genus either 0 or 1, and the genus 1 case happens if and only if every patch of Q has 6 sides and Q has no collapsed edges or vertices.*

Proof. In this case, (6.2.1) becomes

$$\chi(Q) = \frac{1}{2}e + \frac{2}{3}v + \sum_{p \in P} \left(1 - \frac{|p|}{6}\right). \qquad (6.2.2)$$

The terms on the right side of (6.2.2) are all nonnegative, so if any of them are positive, $\chi(Q)$ must be 2. The terms on the right side are all zero if and only if $e = v = 0$ and $|p| = 6$ for all $p \in P$, so the theorem follows. $\quad\square$

Finally, $n = \infty$ gives:

Corollary 6.2.4 (Counting formula for $n = \infty$). *Let Q be a finite quilt with s seams, e collapsed edges, and v collapsed vertices. Then*

$$\chi(Q) = -\frac{1}{6}s + \frac{1}{2}e + \frac{2}{3}v + |P|, \qquad (6.2.3)$$

where $|P|$ is the number of patches in Q. $\quad\square$

Remark 6.2.5. Note that if we fix the number of seams in a modular quilt Q, (6.2.3) provides an upper bound to the possible number of patches of Q and a lower bound to the possible value of $\chi(Q)$. If in addition, we only consider those Q whose patches have order ≤ 6, then only collapsed edges and vertices and patches with order ≤ 5 make a non-zero contribution to the right-hand side of (6.2.2). It follows that, for modular quilts Q whose patches have at most 6 sides, there are only finitely many possible distributions of sets of patch orders $\neq 6$ and numbers of collapsed edges and vertices. In particular, there are only finitely many modular quilts whose patches have order ≤ 5.

Remark 6.2.6. Again, we note that all of the results in this section may be obtained using the Riemann-Hurwitz formula, if the reader prefers that result to the combinatorial Gauss-Bonnet theorem.

7. Classical interpretations of quilts

In this chapter, we show how to use quilts to obtain results on more standard topics in central extension theory and low-dimensional topology. Specifically, we recover a theorem of Conway, Coxeter, and Shephard [14] (Sect. 7.1), we obtain a result on Schur multipliers of quotients of triangle groups (Sect. 7.2), and we interpret the results of Sect. 6.1 in terms of Seifert fibered 3-manifolds (Sect. 7.3).

7.1 A theorem of Conway, Coxeter, and Shephard

Let G be a finitely presented group, let Z be a central subgroup of G, and suppose we are given a finite generating set for Z, along with a way of expressing each generator of Z as a word in the generators of G. In this setting, Conway, Coxeter, and Shephard [14] give an elegant geometric method for explicitly computing a presentation for Z. In this section, we recover one of their main applications, namely, the computation of the orders of various central extensions of polyhedral groups.

Recall that the triangle group $\overline{\Sigma} = (m_1, m_2, n)$ is finite if and only if $(m_1, m_2, n) = (2, 2, n)$, $(2, 3, 3)$, $(2, 3, 4)$, or $(2, 3, 5)$, up to permutation of m_1, m_2, and n. We have the following theorem, which is equivalent to (4.14) of Conway, Coxeter, and Shephard [14], though in slightly different notation.

Theorem 7.1.1. *Let* $\Sigma = \left\langle \frac{p_1}{m_1}, \frac{p_2}{m_2}, \frac{q}{n} \right\rangle$ *be a Seifert group whose projection* $\overline{\Sigma}$ *mod* $\langle Z \rangle$ *has finite order* s. *If* $e(\Sigma) \neq 0$, *then the order of* Σ *is* $s^2 e(\Sigma)$, *and if* $e(\Sigma) = 0$, *then* Σ *is infinite.*

Note that, as mentioned in Conway, Coxeter, and Shephard [14], this theorem contains all prior results on such central extensions of polyhedral groups. See Coxeter and Moser [22, §6.5–6.7] for an overview.

Proof. We first observe that the order of Σ is precisely the order of the quilt Q of the trivial subgroup of Σ. Furthermore, from the relationship between a quilt and its modular structure (Prop. 3.3.11), we know that the modular structure $\overline{Q}$ of Q is precisely the quilt of the trivial subgroup of $\overline{\Sigma}$; and from the quilt covering theorem (Thm. 3.3.25), we know that Q is either the lift of $\overline{Q}$ with the largest possible modulus, or the lift with modulus 0, if such a lift exists.

Now, since the representation of $\overline{\Sigma}$ induced by $\overline{Q}$ is the regular representation (representation of $\overline{\Sigma}$ by right multiplication on its elements), it is fixed-point free, which means that the global collapsing index of $\overline{Q}$ is 1. It then follows from the lifting theorem (Thm. 6.1.7) that if $e(\Sigma) \neq 0$, then no modulus 0 lift of Q exists and the largest possible modulus M of a lift is $M = se(\Sigma)$, and if $e(\Sigma) = 0$, then a lift exists for every modulus, including $M = 0$. Since the order of Σ is sM for $M > 0$ and ∞ for $M = 0$, the theorem follows. $\square$

It is no coincidence that we can recover this theorem of Conway, Coxeter, and Shephard, for in situation at hand, the annotated 2-complex they use to analyze central extensions (called the *Schreier complex*) is precisely the dual of the quilt complex. This duality is also the reason their method involves the homotopy group π_2 and ours involves the homology group H_0. In general, the main difference between the two approaches is that they consider groups other than triangle groups and Seifert groups, but restrict themselves to the regular representation, whereas we restrict ourselves to Seifert groups, but consider arbitrary transitive permutation representations/subgroups of such groups. It may be of some interest to combine the two approaches; see Sect. 13.7 for some ideas.

Let Σ be a Seifert group whose projection $\overline{\Sigma}$ mod $\langle Z \rangle$ is infinite. It is natural to ask if the proof of Thm. 7.1.1 can be used to calculate the order of a central extension of a *quotient* of Σ. (Compare, for instance, example (4.19) from Conway, Coxeter, and Shephard [14].) However, as we will see in Sect. 7.2, applying the proof of Thm. 7.1.1 does *not* produce the desired result, essentially because when the modular structure in question has nonzero genus, the flow solutions constructed by the lifting theorem are no long guaranteed to be *equivariant*. See Sects. 7.2 and 13.7 for an explanation of equivariance, and see Sect. 13.7 for some ideas for further research.

7.2 Central extensions of triangle group quotients

We present the results in this section for two reasons. On the one hand, we show how quilts can be used to construct central extensions of quotients of infinite triangle groups. On the other hand, we also show that if Σ is a Seifert group, Γ is a normal subgroup of $\overline{\Sigma}$ with modular quilt Q, and $\tilde{Q}$ is a quilt with modular structure Q, then $\tilde{Q}$ does *not* necessarily represent a central extension of $\overline{\Sigma}/\Gamma$. The key point is that when Q has genus zero, as in Sect. 7.1, any flow solution coming from a uniform inflow specification actually represents a normal subgroup of Σ, but when Q has higher genus, this can be quite far from the case.

We also remark that the following theorem was inspired by a question of I. Dolgachev, even though the theorem does not quite answer the original question. For a brief explanation, see Rem. 7.2.4.

Theorem 7.2.1. *Let m_1, m_2, and n be pairwise relatively prime, with $\overline{\Sigma} = (m_1, m_2, n)$ infinite (and therefore hyperbolic), and let M be a positive integer. There exists a quotient G of $\overline{\Sigma}$ such that G has a perfect cyclic central extension with kernel of order M.*

For the reader familiar with the theory of the Schur multiplier of a perfect group (see, for instance, Aschbacher [2]), we note that Thm. 7.2.1 implies:

Corollary 7.2.2. *Let m_1, m_2, and n be pairwise relatively prime, with $\overline{\Sigma} = (m_1, m_2, n)$ infinite, and let M be a positive integer. There exists a quotient G of $\overline{\Sigma}$ whose Schur multiplier contains $\mathbf{Z}/M$ as a subgroup (resp. has $\mathbf{Z}/M$ as a quotient).* $\square$

Proof of Thm. 7.2.1. Recall that since m_1, m_2, and n are pairwise relatively prime, there exists a (unique) perfect Seifert group $\Sigma = \left\langle \frac{p_1}{m_1}, \frac{p_2}{m_2}, \frac{q}{n} \right\rangle$, with $|e(\Sigma)| = \frac{1}{m_1 m_2 n}$ (Thm. 2.3.6). We may construct G and the desired central extension $\widetilde{G}$ by finding a normal subgroup $\Delta \lhd \Sigma$ of modulus M that produces the following commutative diagram of exact sequences:

$$
\begin{array}{ccccccccc}
 & & 1 & & 1 & & 1 & & \\
 & & \downarrow & & \downarrow & & \downarrow & & \\
1 & \longrightarrow & \langle Z^M \rangle & \longrightarrow & \langle Z \rangle & \longrightarrow & \mathbf{Z}/M & \longrightarrow & 1 \\
 & & \downarrow & & \downarrow & & \downarrow & & \\
1 & \longrightarrow & \Delta & \longrightarrow & \Sigma & \longrightarrow & \widetilde{G} & \longrightarrow & 1 \\
 & & \downarrow & & \downarrow & & \downarrow & & \\
1 & \longrightarrow & \Gamma & \longrightarrow & \overline{\Sigma} & \longrightarrow & G & \longrightarrow & 1 \\
 & & \downarrow & & \downarrow & & \downarrow & & \\
 & & 1 & & 1 & & 1 & &
\end{array}
\qquad (7.2.1)
$$

Note that if there exist G, $\widetilde{G}$, and so on, such that (7.2.1) commutes, then $\widetilde{G}$ and G are perfect, since they are quotients of Σ.

Now, as is well-known (Thm. 2.2.5), we may choose a torsion-free subgroup Γ_0 of finite index in $\overline{\Sigma}$. Since the abelianization of Γ_0 is nontrivial free abelian (Thm. 2.2.5), let Γ_1 be the kernel of a homomorphism from Γ_0 onto the cyclic group of order $m_1 m_2 n M$, let Q be the modular quilt of Γ_1, and let s be the number of seams in Q (that is, let s be the index of Γ_1 in $\overline{\Sigma}$). Since $s = m_1 m_2 n M$, M divides $se(\Sigma)$, and since Γ_1 is torsion-free, Q has collapsing index 1. The quilt lifting theorem (Thm. 6.1.7) therefore implies that there exists a quilt $\widetilde{Q}$ with modulus M and modular structure Q.

Let Δ_1 be the stabilizer of an element of the representation of Σ induced by $\widetilde{Q}$. The key point now is that *even though Γ_1 is normal in $\overline{\Sigma}$, Σ/Δ_1 is not* necessarily a central extension of $\overline{\Sigma}/\Gamma_1$, since Δ_1 is *not* necessarily normal in Σ. Instead, let Δ be the intersection of the (finitely many) conjugates of Δ_1. Then Δ has finite index in Σ, and since $\langle Z \rangle$ is central, the modulus of Δ is still M. We therefore arrive at the situation shown in (7.2.1), and the theorem follows.   □

Remark 7.2.3. Consider again the following situation from the proof of Thm. 7.2.1: Let Σ be a Seifert group with infinite projection $\overline{\Sigma}$, and let Γ be a torsion-free normal subgroup of finite index s in $\overline{\Sigma}$ with modular quilt Q. Furthermore, assume that Σ is geometric (Defn. 2.3.3) and $e(\Sigma) \neq 0$. Now, suppose M divides $se(\Sigma)$. From Thm. 6.1.7, we know that there exists a quilt $\widetilde{Q}$ with modulus M and modular structure Q. (In fact, from Thm. 6.1.10, we know that there exist M^{2g} distinct such quilts, where g is the genus of Q.) We may then ask: if Δ is the stabilizer of an element of the representation of Σ induced by $\widetilde{Q}$, is Δ normal in Σ?

Now, if the answer to this question were always "yes," then any quotient of (say) the triangle group (2 3 7) of order s would have a Schur multiplier of order divisible by $se(\Sigma) = s/42$. This is most certainly not always the case, and in fact, seems to be almost *never* true. On the other hand, if the arrow flow of the corresponding quilt $\widetilde{Q}$ were *equivariant*, that is, invariant up to homology under the automorphism group $\overline{\Sigma}/\Gamma$ of Q, then Δ would indeed be normal (using the centrality of $\langle Z \rangle$). It must therefore be the case that most arrow flow solutions of a quilt inflow specification, even for a quilt inflow specification on the modular quilt of a normal subgroup of $\overline{\Sigma}$, are not equivariant.

It would be interesting to develop a suitable "equivariant quilt theory" as a method for constructing central extensions of finite quotients of triangle groups. See Ques. 13.6.2 and 13.6.3 for some possibilities for further work.

Remark 7.2.4. We remark that Dolgachev originally asked the following question: Let $\overline{\Sigma}$ be a perfect triangle group (2 3 n). Is there some constant M such that M is larger than the order of the Schur multiplier of any *simple* quotient of $\overline{\Sigma}$? Theorem 7.2.1 gives a partial answer to Dolgachev's question, in that the answer is "no" if the word "simple" is removed. In fact, Lucchini, Tamburini, and Wilson [52] have recently shown that for m and q sufficiently large, $L_m(q)$ is a quotient of (2 3 7). Since the Schur multiplier of $L_m(q)$ has order $\geq \gcd(m, q - 1)$, the answer to Dolgachev's question as stated is also "no."

7.3 Quilts and Seifert fibered 3-manifolds

Finally, in this section, we interpret quilts in terms of the theory of *Seifert fibered 3-manifolds*. For the reader familiar with other aspects of low-dimensional topology, the very brief and informal summary of Seifert fibered 3-manifolds given below may be sufficient, but for more details, we recommend Seifert [79]. See also Orlik [69, Chap. 1, 5] and Scott [78, §1–3].

A Seifert fibered 3-manifold is a "fiber bundle" with base space a compact surface and fiber S^1, such that a neighborhood of any fiber is either

1. $\mathbf{D}^2 \times S^1$, where $\mathbf{D}^2$ denotes the open 2-disc; or
2. $\mathbf{D}^2 \times [0,1]$ with its two ends identified by a p/m twist, where p and m are relatively prime.

Fibers whose neighborhoods are of the second type are called *singular fibers*, and their neighborhoods are called p/m *singular fiber neighborhoods*. For example, Fig. 7.3.1 shows a singular fiber (solid line) and a non-singular fiber (dashed line) inside the 2/5 singular fiber neighborhood obtained by identifying the top and bottom of the cylinder.

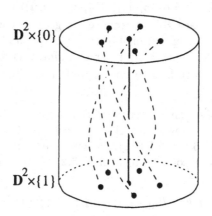

Fig. 7.3.1. A 2/5 singular fiber neighborhood

After "drilling out" all of the (finitely many) singular fiber neighborhoods, or drilling out at least one non-singular neighborhood if there are no singular fibers, we are left with a fiber bundle (in the usual sense) over a surface with boundary. Such a fiber bundle is always trivial, so the Seifert fibered 3-manifold with which we began is determined only by the way we glue the drilled fibers back in. Careful consideration of this "filling" process shows that we may therefore obtain any Seifert fibered 3-manifold by filling fiber neighborhood holes in a trivial bundle over a surface with boundary. We may therefore represent any orientable Seifert fibered 3-manifold with orientable base space by the symbol

$$\langle g | (p_1, m_1), \ldots, (p_t, m_t) \rangle, \qquad (7.3.1)$$

which denotes the space obtained by taking the trivial bundle over a surface of genus g, drilling out t neighborhoods, and filling them with neighborhoods of type (p_i, m_i) $(\gcd(p_i, m_i) = 1)$.

Seifert fibered 3-manifolds are related to quilts by the following theorem.

Theorem 7.3.1 (Seifert). *The fundamental group of the Seifert fibered 3-manifold $\langle 0|(s,1),(p_1,m_1),(p_2,m_2),(q,n)\rangle$ is isomorphic to the Seifert group $\left\langle \frac{p_1}{m_1}, \frac{p_2}{m_2}, \frac{q}{n} \right\rangle_s$.* □

We therefore also define the Euler number of a Seifert fibered 3-manifold to be the Euler number of its fundamental group. Note that all Seifert groups obtained from Thm. 7.3.1 are *geometric* in the sense of Defn. 2.3.3.

From the viewpoint of Seifert fibered 3-manifolds, quilt theory classifies the possible covers of a given Seifert fibered 3-manifold. More specifically, from the theory of covering spaces, we see that Prob. 6.1.2 is equivalent to the following problem.

Problem 7.3.2. Let $E = \langle 0|(p_1,m_1),(p_2,m_2),(q,n)\rangle$ be a Seifert fibered 3-manifold with base space B. Suppose we are given a covering $\widetilde{B}$ of B that is ramified only over the exceptional points of B (points beneath exceptional fibers) and M either a positive integer or infinity. Is there an unramified covering $f : \widetilde{E} \to E$ with base $\widetilde{B}$ and degree M on fibers? If so, how many are there?

Of course, *a fortiori*, an answer to Prob. 7.3.2 solves the question of which coverings of E project down to a given covering on the base space. In any case, since Thms. 6.1.6 and 6.1.10 give a complete solution to Prob. 6.1.2, we also obtain a complete solution to Prob. 7.3.2. We describe one case with a relatively clean statement.

Theorem 7.3.3. *Let $E = \langle 0|(p_1,m_1),(p_2,m_2),(q,n)\rangle$ be a Seifert fibered 3-manifold with base space B. Suppose $\widetilde{B} \to B$ is a covering of degree s that is ramified only over the exceptional points of B (points beneath singular fibers), and suppose $\widetilde{B}$ is maximally ramified at each such point. (That is, suppose that each point of $\widetilde{B}$ mapping to a (p,m) point has ramification of degree m.) Then:*

1. *For every integer $M > 0$, there exists an unramified cover $f : \widetilde{E} \to E$ projecting down to $\widetilde{B} \to B$ with degree M on fibers if and only if M divides $se(E)$, where $e(E)$ is the Euler number of E.*
2. *Furthermore, if such a covering exists, there are precisely M^{2g} distinct such coverings, where g is the genus of $\widetilde{B}$.* □

Remark 7.3.4. We remark that in terms of algebraic surfaces, Seifert fibered 3-manifolds (and therefore, quilts) may be used to study complex line bundles over complex curves; see Neumann and Raymond [61, Sect. 2] for an overview. In fact, this was the original motivation for the question of Dolgachev mentioned in Rem. 7.2.4. See also Ques. 13.7.2.

Part II

The structure problem

8. Presentations and the structure problem

Informally, the *structure problem* for Norton systems may be stated as follows.

> *To what extent does a Norton system for a (finite) group G determine the structure of G?*

In this chapter, we restate this problem rigorously in terms of presentations, and use the geometry and combinatorics of quilts to investigate the problem. We begin by defining the *group of a quilt* (Sect. 8.1). Now, the presentation at the heart of this definition is infinite, so to work with examples, we need a method of reducing it to a managable finite presentation. We discuss several such methods. The simplest and most useful reduction method is the *polyhedral presentation* (Sect. 8.2), which applies to certain quilts of genus 0. A slightly modified version of the polyhedral presentation serves to cover the general genus 0 case (Sect. 8.3). In fact, if we add relations to account for the topology of the quilt, we can expand these methods to the full general case (Sect. 8.4).

All quilts in this chapter are $\mathbf{B}_3$-quilts, and all modular quilts are $\mathbf{PSL}_2(\mathbf{Z})$-modular quilts.

8.1 The group of a quilt

We first define the group of a quilt, and explain the reasoning behind our definition later. The following definition is due to Conway.

Definition 8.1.1. Let Q be a quilt, and recall (Thm. 3.3.20) that there is a unique conjugacy class of subgroups of $\mathbf{B}_3$ associated with Q. We define the *group of Q* to be the group

$$G(Q) = G(\Delta) = \langle \alpha, \beta \,|\, (\alpha, \beta)\pi = (\alpha, \beta), \forall \pi \in \Delta \rangle, \qquad (8.1.1)$$

where Δ is any subgroup in the conjugacy class represented by Q, and $(\alpha, \beta)\pi$ is evaluated in the free group on α and β.

We first verify that Defn. 8.1.1 makes sense.

Theorem 8.1.2. *The group defined by presentation (8.1.1) does not depend on the choice of Δ.*

Proof. It is enough to show that $G(\Delta)$ is isomorphic to both $G(L^{-i}\Delta L^i)$ and $G(R^{-i}\Delta R^i)$. We show only the first case, since the second is essentially the same. Now, by definition,

$$G(\Delta) = \langle \alpha, \beta \,|\, (\alpha, \beta)\pi = (\alpha, \beta), \, \forall \pi \in \Delta \rangle, \tag{8.1.2}$$

$$G(L^{-i}\Delta L^i) = \langle \gamma, \delta \,|\, (\gamma, \delta)\pi = (\gamma, \delta), \, \forall \pi \in L^{-i}\Delta L^i \rangle. \tag{8.1.3}$$

We just have to show that (8.1.2) and (8.1.3) are isomorphic to the group presented by

$$\langle \ \alpha, \beta, \gamma, \delta \ | \tag{8.1.4}$$

$$\alpha = \gamma, \ \delta = \alpha^i \beta, \ \beta = \gamma^{-i}\delta; \tag{8.1.4a}$$

$$(\alpha, \beta)\pi = (\alpha, \beta), \, \forall \pi \in \Delta; \tag{8.1.4b}$$

$$(\gamma, \delta)\pi = (\gamma, \delta), \, \forall \pi \in L^{-i}\Delta L^i \rangle. \tag{8.1.4c}$$

To get from (8.1.2) to (8.1.4), we first add the redundant generators γ and δ and the relations in (8.1.4a). Since $(\alpha, \beta)L^i = (\gamma, \delta)$ follows, we can then derive (8.1.4c) from (8.1.4b), obtaining (8.1.4). Obtaining (8.1.4) from (8.1.3) is entirely analogous, and the theorem follows.   □

Corollary 8.1.3. *If Q_1 covers Q_2, then $G(Q_2)$ is a quotient of $G(Q_1)$.*

Proof. Since Q_1 covers Q_2, some $\Delta_1 \leq \mathbf{B}_3$ in the conjugacy class associated with Q_1 is a subgroup of some $\Delta_2 \leq \mathbf{B}_3$ in the conjugacy class associated with Q_2. By definition, $G(\Delta_2)$ is $G(\Delta_1)$ with extra relations, and the corollary follows.   □

The next theorem justifies Defn. 8.1.1 by showing that if Q is the quilt of a Norton system N, then $G(Q)$ is universal among groups that have a Norton system with quilt Q.

Theorem 8.1.4. *Let Q be the quilt of a Norton system N for a group G. There exists a surjective homomorphism from $G(Q)$ to G. Furthermore, Q is a quilt for $G(Q)$.*

Note that the last statement is certainly not true for arbitrary Q, as it is easy to make a non-trivial quilt Q such that $G(Q)$ is trivial.

Proof. Let (α, β) be a pair in N, and let Δ be its stabilizer. Then

$$G(Q) = \langle \gamma, \delta \,|\, (\gamma, \delta)\pi = (\gamma, \delta), \, \forall \pi \in \Delta \rangle. \tag{8.1.5}$$

However, since Δ stabilizes (α, β), all of the relations in presentation (8.1.5) are respected by the map $\varphi : G(Q) \to G$ defined by $\varphi(\gamma) = \alpha$ and $\varphi(\delta) = \beta$. Therefore, φ is a surjective homomorphism from $G(Q)$ to G. As for the second statement, it is clear from the definition of $G(Q)$ that Δ is contained in the stabilizer of (γ, δ), where γ and δ are again the standard generators of $G(Q)$.

Therefore, Q at least covers the quilt of $N(\gamma, \delta)$. However, by Thm. 4.2.5 and the first statement of this theorem, the quilt of $N(\gamma, \delta)$ also covers Q, the quilt of $N(\alpha, \beta)$, and the theorem follows.   □

Definition 8.1.5. Let Q be a quilt. The *standard Norton system* of $G(Q)$ is defined, in the notation of Defn. 8.1.1, to be $N(\alpha, \beta)$. (From an argument similar to the proof of Thm. 8.1.2, this does not depend on the choice of Δ.) A quilt Q is called *snug* if it is the quilt of the standard Norton system of its group. (The name is meant to suggest that the quilt Q fits G snugly.) In other words, because of Thm. 8.1.4, Q is snug if and only if it is the quilt of a Norton system of some group.

It is an open and apparently difficult problem to characterize when a quilt is snug. This problem is therefore the focus of most of the rest of this monograph.

Remark 8.1.6. Again using the notation of Defn. 8.1.1, let Π be a set of generators of Δ. Clearly, (8.1.1) can be reduced to

$$G(Q) = \langle \alpha, \beta \,|\, (\alpha, \beta)\pi = (\alpha, \beta), \, \forall \pi \in \Pi \rangle . \qquad (8.1.6)$$

If Q is finite, then Δ has finite index in $\mathbf{B}_3$, and the Reidemeister-Schreier method (see App. A) implies that Δ is finitely generated, and therefore, that $G(Q)$ is finitely presented. In practice, however, using Reidemeister-Schreier directly is unwieldy, and produces awkward presentations, so in the rest of this chapter, we consider other methods of reducing quilt presentations to finite presentations.

8.2 The polyhedral presentation

In this section, we show that if a quilt Q has genus 0, and meets a few other simple conditions, we can reduce the presentation of $G(Q)$ to a concise presentation known as the *polyhedral presentation*. We begin with two definitions.

Definition 8.2.1. We say that a quilt Q with s seams is *collapse-free* if Q has no collapsed edges or vertices, all patches of Q are unramified, and the modulus of Q is $s/6$. (Note that, from Cor. 6.1.8, $s/6$ is the largest possible modulus of a quilt with no ramified patches and the modular structure of Q.)

Notation 8.2.2. For the rest of this chapter, if Q is a collapse-free quilt, we do not consider Q in terms of its usual structure as a 2-complex. Instead, we consider Q to have the 2-complex structure obtained by taking vertices (in the sense of quilts) as 0-cells, edges (again in the sense of quilts) as 1-cells, and patches as 2-cells.

Definition 8.2.3. Let Q be a finite collapse-free quilt, and let Ξ be a tree such that:

1. Ξ has a node for every patch of Q.
2. For every edge of Ξ between nodes representing patches P_1 and P_2, there exists an edge of Q at which P_1 attaches to P_2.

Furthermore, let X be the subspace of Q such that:

1. For every node of Ξ, X contains the interior of the corresponding patch of Q.
2. For every edge of Ξ, X contains the interior of the corresponding edge of Q, joined to the two patches it touches.

The subspace X is then said to be a *spanning tree of polygons* for Q.

Remark 8.2.4. It is important to note that by construction, X contains no vertices (0-cells), just edges and patches. More precisely, if Q has P patches, then X consists of P patches and $P - 1$ edges.

A spanning tree of polygons for Q may be conveniently indicated on Q itself. For example, consider the quilt in Fig. 8.2.1, to which we will return throughout this section. Note that we have omitted the dashes and arrow flows in Fig. 8.2.1, except to indicate that Q has no patch inflows. (In fact, from Thm. 6.1.10, we know that Q is actually determined by this simplified diagram, once we specify the modulus.)

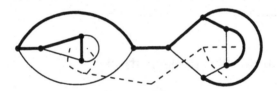

Fig. 8.2.1. Genus 0 quilt, modulus $M = 5$

In the notation of Defn. 8.2.3, the dashed lines in Fig. 8.2.1 represent the graph Ξ, and the spanning tree of polygons X is made from the edges and patches that Ξ touches. Note that from Thm. 2.6.6, every finite collapse-free quilt has a spanning tree of polygons.

Definition 8.2.5. Let Q be a finite collapse-free quilt of genus 0, and let X be a spanning tree of polygons for Q. The *complementary cut* to X is defined to be the subspace $Q - X$ of Q. Note that the complementary cut naturally forms a graph that is a subcomplex of Q, and contains all vertices of Q.

The complementary cut to the spanning tree of our example is indicated by the heavy edges in Fig. 8.2.1. It is useful to consider what happens when we actually apply these cuts. For instance, in our example, the result is shown

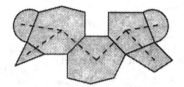

Fig. 8.2.2. Flattened spanning tree

in Fig. 8.2.2. (For clarity, we have shaded in the polygons.) This type of diagram, which is easily made from a spanning tree of polygons, is called a *flattened spanning tree* for Q.

Next, we ask: Given a flattened spanning tree, can we determine the quilt from which it came? In our example, the reader should verify that Q may be recovered from the flattened spanning tree by repeatedly applying the rule that the 1-skeleton of Q is a trivalent graph. The following proposition describes a general case of this phenomenon.

Theorem 8.2.6. *Let Q be a finite collapse-free quilt of genus 0, and let X be a spanning tree of polygons for Q. We can recover the modular structure of Q from the flattened spanning tree.*

Proof. Let $A = Q - X$ be the complementary cut to X. Since $X = Q - A$ is connected and Q is homeomorphic to $\mathbf{S}^2$, by the Jordan curve theorem (Thm. 2.6.11), A cannot contain any circuits. (For another proof, see Thm. 8.4.4.) In other words, A is a disjoint union of trees.

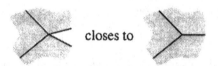

closes to

Fig. 8.2.3. The trivalent rule

So now, consider the cut pictured in Fig. 8.2.3. In this situation, by applying the *trivalent rule*, we can close up the cut in the indicated manner. However, the situation in Fig. 8.2.3 is precisely the situation at a terminal edge of A, so we can undo any cut represented by a terminal edge of A. Since A is a finite disjoint union of finite trees, the proposition follows by pruning terminal edges of A (Thm. 2.6.4). □

We are now ready to prove the main theorem of this section. However, we need a definition before stating our result.

Definition 8.2.7 (The polyhedral procedure). Let Q be a collapse-free finite quilt of genus 0, and let X be a flattened spanning tree of polygons

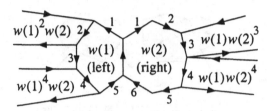

Fig. 8.2.4. Orientations and the labelling rule

for Q. We have the following procedure for labelling X with words in α and β, where α and β are generators of a free group.

1. Orient each polygon of X as either a *left* polygon or a *right* polygon, so that every left polygon only touches right polygons, and vice versa. (This is possible because X is a tree.) We indicate the orientation of a left (resp. right) polygon by directing its edges counterclockwise (resp. clockwise), as shown in Fig. 8.2.4.
2. Pick two adjacent polygons and label the left one α and the right one β.
3. Label the rest of the polygons of X using the following (recursive) rule: Suppose two adjacent polygons P_1 and P_2 are already labelled $w(1)$ and $w(2)$, with P_1 a left polygon and P_2 a right polygon. Paying careful attention to the orientations of P_1 and P_2, number their sides as shown in Fig. 8.2.4, and label any previously unlabelled patch touching side n of P_1 (resp. P_2) with $w(1)^n w(2)$ (resp. $w(1)w(2)^n$).

Now, since X is a tree of polygons, this process is well-determined and labels all of the polygons in X. We may therefore define the *polyhedral presentation* determined by X to be

$$G(X) = \left\langle \alpha, \beta \,\middle|\, 1 = w(i)^{k(i)} \right\rangle, \tag{8.2.1}$$

where i runs over all polygons P_i in X, $w(i)$ is the label of P_i, and $k(i)$ is the number of sides of P_i.

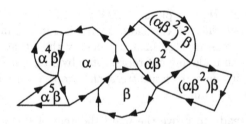

Fig. 8.2.5. Labelled flattened spanning tree

Let Q be the quilt in Fig. 8.2.1, p. 104. Figure 8.2.5 shows the result of performing the polyhedral procedure on the spanning tree of polygons from

Fig. 8.2.1. The corresponding polyhedral presentation is

$$G(X) = \left\langle \alpha, \beta \;\middle|\; \begin{array}{l} 1 = \alpha^7 = \beta^7 = (\alpha^4\beta)^2 = (\alpha^5\beta)^3 \\ = (\alpha\beta^2)^5 = (\alpha\beta^3)^4 = (\alpha\beta^2\alpha\beta^3)^2 \end{array} \right\rangle. \qquad (8.2.2)$$

Now, suppose we want to draw the quilt Q_1 of $N(\alpha, \beta)$, where α and β are the generators of $G(X)$. At first, let us assume that Q_1 is collapse-free. Given this assumption, by applying the multiplication principle (Cor. 4.3.5) and the patch theorem (Cor. 4.3.7), and using the defining relations of $G(X)$, we recover the labelled flattened spanning tree from Fig. 8.2.5. Next, applying the trivalent rule, we see that the labelled spanning tree closes up into a quilt diagram (Q_0, M_0) with the modular structure of Q. Note that from Sect. 4.4, we know that Q_0 at least *covers* Q_1.

We leave it as an exercise for the reader to check that, still assuming Q_0 is collapse-free, applying the flow rules to Q_0 forces the congruence $0 \equiv 5$ (mod M_0). In any case, since Q_0 has genus 0, modulus 5, and the same modular structure as Q, and is also collapse-free, it must actually be equal to Q (Cor. 6.1.11). Therefore, the relations in $G(X)$ imply the defining relations of $G(Q)$, which means that $G(Q)$ is a quotient of $G(X)$. In fact, since $G(X)$ is clearly a quotient of $G(Q)$, $G(X)$ must actually be isomorphic to $G(Q)$.

More generally, a precise version of the above argument yields the following theorem.

Theorem 8.2.8 (Polyhedral presentations). *Let (Q, M) be a collapse-free finite quilt of genus 0. Any polyhedral presentation determined by a spanning tree of polygons for Q is isomorphic to $G(Q)$.*

Proof. Let X be a spanning tree of polygons for (Q, M), let $G(X)$ be the group defined by the corresponding polyhedral presentation (8.2.1), and let (Q_1, M_1) be the quilt of $N(\alpha, \beta)$, where α and β are the generators of $G(X)$. Since the multiplication principle (Cor. 4.3.5) and the patch theorem (Cor. 4.3.7) imply that the relations of (8.2.1) hold in $G(Q)$, $G(Q)$ is a quotient of $G(X)$. Therefore, from Cor. 8.1.3, it suffices to show that Q covers Q_1.

First, we observe that the defining relations of (8.2.1) allow us to recover the flattened spanning tree of polygons for X by repeated applications of the multiplication principle and the patch theorem. (More precisely, each "oriented edge" that indicates the orientation of a patch of X can be considered as shorthand for a basic edge; compare Cor. 4.3.3.) From Thm. 8.2.6, we then see that the modular structure of Q covers the modular structure of Q_1. Let (Q_0, M_1) be the quilt obtained by pulling the arrow flows of Q_1 back to the modular structure of Q (Thm. 3.3.34).

The key assertion is that Q_0 has no ramified patches, which we can see as follows. Let P_0 be a patch of Q_0, let P_1 be the patch of Q_1 that P_0 covers, and for $j = 0, 1$, let the order, inflow, and ramification index of P_j be denoted by k_j, i_j, and n_j, respectively. By construction, we know that any stack element of P_1 has order dividing k_0. The patch theorem then implies that $k_0 = c n_1 k_1$

for some integer c. However, since every quilt morphism is a patch inflow covering (Exer. 6.1.16),

$$i_0 \equiv \frac{k_0}{k_1} i_1 \equiv cn_1 i_1 \equiv 0 \pmod{M_1}, \qquad (8.2.3)$$

since n_1 is the additive order of i_1 (mod M_1). Therefore, P_0 is unramified.

Now, since Q and Q_0 have the same modular structure and no ramified patches, Cor. 6.1.8 implies that M_1 divides $s/6 = M$. Then, since Q has genus 0, Cor. 6.1.11 implies that Q_0 is the reduction of Q mod M_1. Therefore, Q covers Q_0, which in turn covers Q_1, and the theorem follows. □

Remark 8.2.9. It is not hard to see from Thm. 8.2.6 that a polyhedral presentation can be used to recover the modular structure of a quilt Q. The surprising part of Thm. 8.2.8 is that we can get the lift to $\mathbf{B}_3$ "for free", by using the homology results of Sect. 6.1. In other words, we have used our lifting results to show that polyhedral relations like those in (8.2.2) imply the relations $\alpha_M = \alpha$ and $\beta_M = \beta$.

8.3 Other genus zero presentations

As we shall see momentarily, for a quilt Q of genus 0 that is not collapse-free, the conclusion of Thm. 8.2.8 only holds in certain special cases. More generally, however, thinking along the lines of Thm. 8.2.8, we show that the polyhedral presentation can be modified to give a presentation for the group of any quilt of genus 0.

We begin with two cases where Thm. 8.2.8 works as before.

Theorem 8.3.1. *Let Q be a finite quilt of genus 0 with one collapsed edge (resp. one collapsed vertex), and no other collapsing or ramified patches. Let s be the number of seams in Q, and suppose that the modulus of Q is equal to $s/3$ (resp. $s/2$). Then any polyhedral presentation determined by a spanning tree of polygons for Q is isomorphic to the group of Q.*

Proof. Since the modulus of Q is maximal given the modular structure of Q and the assumption that Q is collapse-free except for a single collapsed edge or vertex, the proof is almost the same as that of Thm. 8.2.8; the only difference comes when we are trying to recover the modular structure of Q. The complementary cut is again a tree, with one additional feature; namely, the cut needed to open up the collapsed edge (resp. vertex) of Q must be one of the terminal edges of the complementary cut. However, since any finite tree with at least 3 vertices has at least 2 terminal edges, we can still use the trivalent rule (Fig. 8.2.3, p. 105) to undo the rest of the complementary cut until we are left with the one exceptional edge. At that point, the situation at the exceptional edge looks like Fig. 8.3.1 (resp. 8.3.2), and this last cut can also be undone, as shown. □

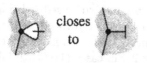
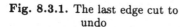
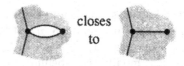

Fig. 8.3.1. The last edge cut to undo

Fig. 8.3.2. The last vertex cut to undo

For a general quilt Q of genus 0, which may have collapsed edges and vertices, ramified patches, and a non-maximal modulus, we essentially need to account for enough collapse to reduce the situation to the situation in Thm. 8.2.8. We need to be able to handle the modulus, collapsed vertices, collapsed edges, and patch ramification, so we discuss these phenomena in that order.

The modulus of the quilt is the easiest to account for, as we can simply add the relations $\alpha_M = \alpha$ and $\beta_M = \beta$. It is useful to note that these relations are often still redundant; for instance, if Q has no patch inflows, and M is maximal for the modular structure of Q, our homology results can be applied to recover the modulus of Q, as before.

The other three types of collapse are handled in the following way. The basic strategy is to cut Q into a "spanning tree of collapsed polygons" from which the modular structure of Q can be recovered, and then determine appropriate relations to recover the flattened spanning tree.

Fig. 8.3.3 shows a quilt of genus 0, and Fig. 8.3.4 shows a flattened spanning tree of collapsed polygons for the quilt in Fig. 8.3.3. Note that in contrast to the process used in Thm. 8.3.1, we have left the collapsed edge and collapsed vertex sealed up in Fig. 8.3.4.

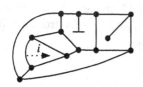
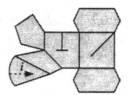

Fig. 8.3.3. Quilt of genus 0

Fig. 8.3.4. Flattened spanning tree of collapsed polygons

At this point, we would like to imitate the polyhedral procedure (Defn. 8.2.7) to obtain a presentation for $G(Q)$. The problem is, for polygons that contain some kind of collapse, we must replace the relation $1 = w(i)^{k(i)}$ in (8.2.1) with more complicated relations. As we will see in Sect. 9.1, in a snug quilt, any patch containing a collapsed edge (resp. vertex) is unramified, and a patch cannot contain more than one collapsed edge (resp. vertex), so we only need to consider the four types of collapsed polygons shown in Fig. 8.3.5.

In the following, we denote the number of sides of a collapsed polygon by p. For example, in Fig. 8.3.5, $p = 4$ for all four cases, even though the patches producing cases (1)–(3) have orders 5, 6, and 7, respectively. Note that we must also modify the labelling rule of Fig. 8.2.4 in cases (1)–(3) to label the collapsed edges and vertices that appear.

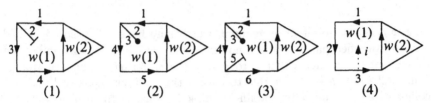

Fig. 8.3.5. The four types of collapsed polygons

In case (1), if side n of the collapsed polygon comes from a collapsed edge ($n = 2$ in Fig. 8.3.5), we use the relations

$$1 = w(1)^{p+1},$$
$$(w(1), w(2))L^n V_2 = (w(1), w(2))L^n Z^{-\frac{1}{2}}, \tag{8.3.1}$$

where $\frac{1}{2}$ is the multiplicative inverse of 2 mod M. Similarly, in case (2), if sides n and $n + 1$ come from a collapsed vertex ($n = 2$ in Fig. 8.3.5), we use the relations

$$1 = w(1)^{p+2},$$
$$(w(1), w(2))L^{n+1} V_1 = (w(1), w(2))L^{n+1} Z^{\frac{1}{3}}, \tag{8.3.2}$$

where $\frac{1}{3}$ is the multiplicative inverse of 3 mod M. In case (3), where sides n_1 and $n_1 + 1$ come from a collapsed vertex and side n_2 comes from a collapsed edge ($n_1 = 2$ and $n_2 = 5$ in Fig. 8.3.5), we use

$$1 = w(1)^{p+3},$$
$$(w(1), w(2))L^{n_1+1} V_1 = (w(1), w(2))L^{n_1+1} Z^{\frac{1}{3}}, \tag{8.3.3}$$
$$(w(1), w(2))L^{n_2} V_2 = (w(1), w(2))L^{n_2} Z^{-\frac{1}{2}}.$$

Finally, in case (4), we simply use

$$(w(1), w(2))L^p = (w(1), w(2))Z^i, \tag{8.3.4}$$

where i is the inflow of the patch.

Note that the above examples depend on the collapsed polygon being on the left. If we consider the mirror images of the four cases of Fig. 8.3.5, in (8.3.1)–(8.3.3) we just replace L with R, and (8.3.4) becomes

$$(w(1), w(2))R^p = (w(1), w(2))Z^{-i}. \tag{8.3.5}$$

(Compare Thm. 3.3.28.)

In any case, we can now define the *collapsed polyhedral procedure* by analogy with the polyhedral procedure, replacing $1 = w(i)^{p(i)}$ with (8.3.1)–(8.3.5), as necessary, and adding the modulus relations $\alpha_M = \alpha$ and $\beta_M = \beta$. The presentation obtained is known as a *collapsed polyhedral presentation*. Imitating the proof of Thm. 8.2.8 then yields the following.

Theorem 8.3.2. *Let Q be a finite quilt of genus 0. Any collapsed polyhedral presentation determined by a spanning tree of collapsed polygons for Q is isomorphic to $G(Q)$.*   □

8.4 Higher genus presentations

In this section, we show how to find a relatively uncomplicated finite presentation for the group of a quilt Q of nonzero genus, by modifying the polyhedral presentation to account for the topology of Q. For simplicity, we give details only in the collapse-free case.

Let Q be a collapse-free finite quilt of genus $g > 0$. First, we note that the definition of a spanning tree of polygons X for Q (Defn. 8.2.3) works just as well for $g > 0$ as it does for $g = 0$. We may therefore make the following definition.

Definition 8.4.1. Let Q be a collapse-free finite quilt of genus $g > 0$ and let X be a spanning tree of polygons for Q. Orient the polygons of X alternately left or right, as in Defn. 8.2.7. We define the *quilt diagram determined by X* to be the unique quilt diagram representative for Q such that:

1. The patches of Q have no inflow anywhere, and
2. The edges of Q in the *interior* of X become basic edges pointed in the direction corresponding to the orientations of the polygons of X.

In particular, note that if Q is snug, and X is a spanning tree of polygons for Q, then the quilt diagram determined by X is precisely the canonically labelled quilt diagram (see Sect. 4.3) obtained by labelling the patches of Q with the elements specified by the labels of the polygons of X from the polyhedral procedure (Defn. 8.2.7).

Now, when Q has nonzero genus, the principal new difficulty to extending the results of Sect. 8.2 is that the complementary cut to a spanning tree of polygons (Defn. 8.2.5) is no longer a disjoint union of trees. We must therefore modify Defn. 8.2.7 in the following manner.

Definition 8.4.2 (The augmented polyhedral procedure). Let Q be a collapse-free finite quilt with modulus M and genus $g > 0$, let X be a spanning tree of polygons for Q, and let A be the complementary cut to X. We have the following procedure.

1. Using the polyhedral procedure (Defn. 8.2.7), label the polygons of X with words in α and β.
2. Choose a set A_0 of 1-cells in A such that $A - A_0$ is a disjoint union of trees. We call A_0 the set of *forced edges*, and we call $A - A_0$ the *reduced complementary cut*.

Note that each forced edge $a \in A_0$ identifies two boundary edges of X in Q. The fact that these boundary edges are identified may be forced by a certain relation in α and β. Specifically, let Q' be the quilt diagram determined by X. For every forced edge a, we define the *edge-forcing relation* $R(a)$ by one of the three cases described below. In each case, for the polygon labels v_i, w_i, $i = 1, 2$, let B_i be the seam labelled $(v_i, 0, w_i)$ in Q', and let $(B_i, 0)$ be the Z-orbit representative of B_i in the representation induced by the quilt diagram Q'.

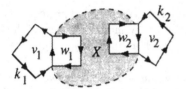
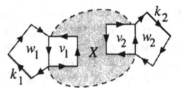

Fig. 8.4.1. Left-left identification, $k_1 = k_2 = 3$

Fig. 8.4.2. Right-right identification, $k_1 = k_2 = 2$

1. Using the edge numbering rule from Fig. 8.2.4, p. 106, suppose a identifes the edges numbered k_1 and k_2 of two left polygons, as shown in Fig. 8.4.1. Then, since a identifies the indicated edges in Q', we must have $(B_1, 0)L^{k_1} = (B_2, 0)L^{k_2}V_2Z^j$ for a unique integer j (mod M). (Note that j is well-defined precisely because we have chosen a specific quilt diagram representative Q'.) We may therefore define $R(a)$ to be the relation

$$(v_1, w_1)L^{k_1} = (v_2, w_2)L^{k_2}V_2Z^j. \tag{8.4.1}$$

(Note that the V_2 appears in (8.4.1) precisely because the identified edges have conflicting orientations.)
2. Similarly, if a identifies the edges numbered k_1 and k_2 of two right polygons, as shown in Fig. 8.4.2, we must have $(B_1, 0)R^{k_1} = (B_2, 0)R^{k_2}V_2Z^j$ for a unique integer j (mod M). In that case, we define $R(a)$ to be the relation

$$(v_1, w_1)R^{k_1} = (v_2, w_2)R^{k_2}V_2Z^j. \tag{8.4.2}$$

3. Finally, if a identifies edge k_1 of a left polygon with edge k_2 of a right polygon, as shown in Fig. 8.4.3, we have $(B_1, 0)L^{k_1} = (B_2, 0)R^{k_2}Z^j$ for some unique integer j (mod M), and we define $R(a)$ to be the relation

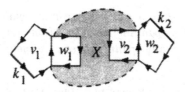

Fig. 8.4.3. Left-right identification, $k_1 = 3$, $k_2 = 2$

$$(v_1, w_1)L^{k_1} = (v_2, w_2)R^{k_2}Z^j. \tag{8.4.3}$$

We may therefore define the *augmented polyhedral presentation* determined by X and A_0 to be

$$G(X, A_0) = \left\langle \alpha, \beta \,\middle|\, 1 = w(i)^{k(i)}, R(a) \right\rangle, \tag{8.4.4}$$

where i runs over all polygons P_i in X, $w(i)$ is the label of P_i, $k(i)$ is the number of sides of P_i, a runs over all forced edges, and $R(a)$ is the edge-forcing relation of a.

We also often use the following alternate form of (8.4.1)–(8.4.3).

Definition 8.4.3. In the notation and terminology of Defn. 8.4.2, suppose we want to identify the edges k_1 and k_2 of the spanning tree X. For $i = 1, 2$, let π_i be an element of $\mathbf{B}_3$ that may be applied to "walk" from the basepoint (the edge between the patches labelled α and β) to k_i, travelling only inside X. In other words, if A is the seam in X with the patch α on the left and the patch β on the right, and K_i is the seam of k_i with no arrow flows on it (thinking of k_i as a basic edge), let $\pi_i \in \mathbf{B}_3$ be an element such that we can deduce $A\pi_i = K_i$ purely from knowledge of the spanning tree X. Then (8.4.1) and (8.4.2) may be expressed in the form

$$(\alpha, \beta)\pi_1 = (\alpha, \beta)\pi_2 V_2 Z^{-j} \tag{8.4.5}$$

for some integer j, and (8.4.3) may be expressed in the form

$$(\alpha, \beta)\pi_1 = (\alpha, \beta)\pi_2 Z^{-j} \tag{8.4.6}$$

for some integer j.

Therefore, in the above notation, for each forced edge a, we may define the *edge-forcing element* $\varphi(a)$ to be $\pi_1 Z^j V_2^{-1} \pi_2^{-1}$ in the case of (8.4.5), and $\pi_1 Z^j \pi_2^{-1}$ in the case of (8.4.6). In this notation, the edge-forcing relation $R(a)$ can be written as $(\alpha, \beta) = (\alpha, \beta)\varphi(a)$. Note that by definition, the edge-forcing relations are precisely the relations needed to force the edge-forcing elements to be in the stabilizer of the basepoint.

See Exmp. 8.4.9 for some examples of edge-forcing elements.

Now, the reader may have noticed that the size of the forced edge set in Defn. 8.4.2 is unspecified, though obviously finite. However, we have the following precise result.

Theorem 8.4.4. *Let Q be a collapse-free finite quilt of genus $g > 0$. Any set of forced edges for a spanning tree of polygons for Q contains at least $2g$ edges. Furthermore, for any spanning tree of polygons, there exists a set of forced edges of size $2g$.*

Proof. Let X be a spanning tree of polygons for Q, and let A be the complementary cut to X. Note that A is connected, since it is the continuous image in Q of the boundary of X, which is a connected set.

Let the number of vertices, edges, and patches of Q be denoted by V, E, and P, respectively; and let the number of vertices and edges of A be denoted by v_0 and e_0, respectively. From Euler's Formula (Thm. 2.6.10), $V - E + P = 2 - 2g$. Therefore, since X contains $P - 1$ edges, we see that $v_0 = V$, $e_0 = E - (P - 1)$, and

$$v_0 - e_0 = V - E + P - 1 = 1 - 2g. \tag{8.4.7}$$

Recall (Thm. 2.6.2) that a graph with v vertices, e edges, and c components is a disjoint union of trees if and only if $v = e + c$. It follows that, to remove edges to reduce A to a disjoint union of trees, we must reduce the number of edges in A until it is strictly less than v_0. Therefore, since (8.4.7) implies that $e_0 = v_0 + 2g - 1$, any set of forced edges for X must contain at least $2g$ edges. Conversely, it is easy to see that if a connected graph A' contains a circuit, then we may remove one of its edges and still have a connected graph. Applying this principle repeatedly to A, we may remove a collection A_0 of $2g$ edges from A to obtain a connected graph $A - A_0$ with $v_0 - 1$ edges. From Thm. 2.6.2, $A - A_0$ must be a tree, which means that A_0 is a forced set of edges of size $2g$. The theorem follows.   □

Remark 8.4.5. Note that if $g = 0$, the above proof shows that the initial complementary cut is already a tree, providing an alternative to the use of the Jordan curve theorem in the proof of Thm. 8.2.6.

In any case, regardless of the size of A_0, we have the following theorem.

Theorem 8.4.6 (Augmented polyhedral presentations). *Let Q be a collapse-free finite quilt of genus $g > 0$. Any augmented polyhedral presentation determined by a spanning tree of polygons for Q is isomorphic to $G(Q)$.*

Proof. Let X be a spanning tree of polygons for Q, let A be the complementary cut to X, let A_0 be a forced edge set for A, let $G(X, A_0)$ be the group defined by the augmented polyhedral presentation determined by X and A_0, and let (Q_1, M_1) be the quilt of $N(\alpha, \beta)$, where α and β are the generators of $G(X, A_0)$. Following the proof of Thm. 8.2.8, it is enough to show that Q covers Q_1.

First, we claim that relations in $G(X, A_0)$ imply that the modular structure of Q_1 is covered by the modular structure of Q. To begin with, we may recover the flattened version of X from the polyhedral relations, as before.

Furthermore, the edge-forcing relations then allow us to recover the identifications along the forced edge set A_0. Then, since the only part of Q left unsealed is the reduced complementary cut $A - A_0$, which is a disjoint union of trees, applying the trivalent rule (Fig. 8.2.3, p. 105), we recover the modular structure of Q. The claim follows.

So now, still following the proof of Thm. 8.2.8, let (Q_0, M_1) be the quilt obtained by pulling back the arrow flows of Q_1 to the modular structure of Q (Thm. 3.3.34). The proof of Thm. 8.2.8 then shows that Q_0 has no ramified patches, and Cor. 6.1.8 then implies that M_1 divides $s/6 = M$. The problem now is that, since Q and Q_0 do not have genus 0, we cannot yet conclude that they are isomorphic. However, let Q' (resp. Q'_0) be the quilt diagram representative for Q (resp. Q_0) determined by X (Defn. 8.4.1). We claim that Q' and Q'_0 have exactly the same arrow flows (mod M_1).

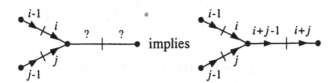

Fig. 8.4.4. The trivalent flow rule

Now, by construction, the arrow flows of Q' and Q'_0 agree everywhere inside X, so it remains only to show that they agree on the complementary cut A. In fact, the edge-forcing relations (8.4.1)–(8.4.3) imply that Q' and Q'_0 agree on the forced edge set A_0, so it remains only to show that Q' and Q'_0 agree on the reduced complementary cut $A - A_0$. However, from the trivalent flow rule (Fig. 8.4.4), we see that the flow on a terminal edge of the "unknown" flows of Q' and Q'_0 is completely determined by the "known" flows. Furthermore, since the reduced complementary cut is a disjoint union of trees, it can be eliminated by pruning terminal edges. Therefore, by repeated application of the trivalent flow rule, the known flows of Q' and Q'_0 determine the flows on the reduced complementary cut. The theorem follows.   □

Remark 8.4.7. We remark that the "trivalent flow rule" argument used to finish the proof of Thm. 8.4.6 also provides an alternative to the use of Cor. 6.1.11 in the proof of Thm. 8.2.8.

Remark 8.4.8. Let Q be a finite modular quilt of genus $g > 0$ with no collapsed vertices or edges, let A_0 be a minimal forced edge set for the complementary cut to a spanning tree of polygons for Q, let s be the number of seams of Q, and let M be an integer dividing $s/6$. Note that on the one hand, A_0 has size $2g$ (Thm. 8.4.4), and on the other hand, there are $(\mathbf{Z}/M)^{2g}$ possible collapse-free quilts with modular structure Q and modulus M (Cor. 6.1.12). It follows that the collapse-free quilts with modular structure Q are freely parameterized by the $2g$ possible choices of a $\mathbf{Z}/M$ parameter j in the edge-forcing relations (8.4.1)–(8.4.3). Indeed, this is one concrete interpretation of the parameters from Cor. 6.1.12.

We end this section with an example.

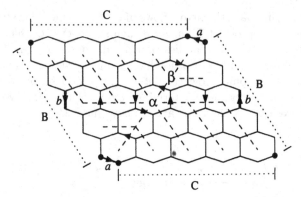

Fig. 8.4.5. $5 \cdot 150^1 (6^{25})$ quilt for $U_3(5)$

Example 8.4.9. Let

$$\alpha = (1\ 2\ 3\ 4\ 5\ 6)(7\ 8\ 9\ 10\ 11\ 12)(13\ 14\ 15\ 16\ 17\ 18)$$
$$(19\ 20\ 21\ 22\ 23\ 24)(25\ 26\ 27\ 28\ 29\ 30)(31\ 32\ 33\ 34\ 35\ 36)$$
$$(37\ 38\ 39)(40\ 41\ 42)(43\ 44\ 45)(46\ 47)(48\ 49)(50)$$

and

$$\beta = (1\ 49\ 33\ 8\ 37\ 21)(2\ 10\ 39\ 23\ 34\ 48)(3\ 16\ 6\ 43\ 30\ 45)$$
$$(4\ 28\ 11\ 20\ 26\ 5)(7\ 24\ 17\ 36\ 31\ 15)(9\ 50\ 22)(12\ 19)$$
$$(13\ 40\ 32\ 27\ 35\ 42)(14\ 47\ 18)(25\ 29\ 46)(38)(41\ 44).$$

A calculation (using, for instance, GAP [77]) shows that $\langle \alpha, \beta \rangle \cong U_3(5)$, the simple group of order 126000, and that the quilt Q of $N(\alpha, \beta)$ has modulus 5 and the modular structure obtained by identifying the borders of the region in Fig. 8.4.5, as shown.

Let X be the spanning tree of polygons indicated by the dashed lines in Fig. 8.4.5. A careful examination of Fig. 8.4.5 shows that the edges a and b are a forced edge set for the complementary cut to X (exercise). As for the edge-forcing relations, in the notation of Defn. 8.4.3, we see that for the forced edge a, we may choose

$$\pi_1 = L^3 R^3 L^3 \qquad\qquad \pi_2 = R^3 L^3, \qquad (8.4.8)$$

and for the forced edge b, we may choose

$$\pi_1 = L^2 R^3 L^3 \qquad\qquad \pi_2 = L^5 R^3 L^3. \qquad (8.4.9)$$

It follows that we may take the edge-forcing elements to be

$$\varphi(a) = L^3 R^3 L^3 Z^j V_2^{-1} L^{-3} R^{-3}, \qquad (8.4.10)$$
$$\varphi(b) = L^2 R^3 L^3 Z^k V_2^{-1} L^{-3} R^{-3} L^{-5}, \qquad (8.4.11)$$

for some integers j and k.

In fact, a calculation in $U_3(5)$ shows that we may choose $j = k = 2$, so by applying the augmented polyhedral procedure, and adjoining relations to force the modulus of Q to be 5, we see that

$$G(Q) \cong \langle\, \alpha, \beta \mid w(i)^6 = 1,\ \alpha_5 = \alpha,\ \beta_5 = \beta,$$
$$(\alpha, \beta) = (\alpha, \beta) L^3 R^3 L^3 Z^2 V_2^{-1} L^{-3} R^{-3}, \qquad (8.4.12)$$
$$(\alpha, \beta) = (\alpha, \beta) L^2 R^3 L^3 Z^2 V_2^{-1} L^{-3} R^{-3} L^{-5} \,\rangle,$$

where i runs over all 25 patches of X and $w(i)$ is the label of patch i.

Remark 8.4.10. Note that the collapsed polygon techniques from Sect. 8.3 may be combined with the ideas in this section to obtain a presentation for the group of a quilt of arbitrary genus that is not necessarily collapse-free. We leave the details to the interested reader.

Remark 8.4.11. Finally, we remark that the major results of this chapter may be interpreted in terms of finding generating sets of subgroups of $\mathbf{B}_3$. For example, let Δ be a finite index subgroup of $\mathbf{B}_3$ whose quilt is collapse-free with p patches and genus g. Then Defn. 8.4.2 and Thms. 8.4.4 and 8.4.6 imply that there exists a generating set for Δ consisting of p conjugates of powers of L and $2g$ edge-forcing elements.

We can also interpret the results of this chapter in terms of finding generating sets of subgroups of $\mathbf{PSL}_2(\mathbf{Z})$. For details, see App. A.

9. Small snug quilts

In this chapter, we enumerate the snug quilts (Defn. 8.1.5) with at most 12 seams and order less than 144.

9.1 Enumerating snug quilts

To find the snug quilts with s seams, we first enumerate all quilts with s seams, as described in Sect. 5.2. Then, for each such quilt Q, we compute $G(Q)$ using one of the presentations of Chap. 8. (For the groups in this chapter, this was done either by recognizing $G(Q)$ as a well-known group, or by determing $G(Q)$ by computer coset enumeration, using the computational group theory system GAP [77].) Once $G(Q)$ is determined, we then check that Q is the quilt of the standard Norton system of $G(Q)$ (Defn. 8.1.5).

To reduce the number of candidates, we also use two simple rules about snug quilts. The first rule is called the *redundancy principle*. This is not so much a theorem as an observation that, by definition, no two seams in a snug quilt can represent the same Z-orbit. For instance, this means that a patch (represented by, say, α) can have at most one collapsed edge, since the seam involved in each collapsed edge would represent the Z-orbit of $(\alpha_{\frac{1}{2}}, \alpha)$. Similarly, a patch can have at most one collapsed vertex. Therefore, we can eliminate all quilts with a patch having more than one collapsed edge or a patch having more than one collapsed vertex.

Note that "redundancy" may occur in $\mathbf{B}_3$-quilts that are not quilts of Norton systems, as shown by the quilt Q of the subgroup of index 2 in $\mathbf{B}_3$ (Fig. 9.1.1). If Q were snug, both seams of Q would represent the Z-orbit of (α, α), where α would be the stack element for the lone patch of the quilt.

Fig. 9.1.1. A "redundant" quilt, $M = 1$ **Fig. 9.1.2.** A ramified self-touching patch, $M = 5$

The second rule is:

Theorem 9.1.1. *Self-touching patches of a snug quilt are unramified.*

Proof. Let (Q, M) be a snug quilt with a self-touching patch P, and let α be a stack element of P. Since P touches itself, some seam of Q represents the Z-orbit of (α, α_j) for some j mod M, and consequently, $M = \text{size}(\alpha)$. However, from the patch theorem for Norton systems (Cor. 4.3.7), M then divides the inflow of the patch of α, which means that the patch of α is unramified. $\square$

Fig. 9.1.2 gives an example showing that Thm. 9.1.1 does not apply to non-snug quilts. Specifically, the lone patch of this quilt touches itself and has ramification index 5, so this quilt cannot be snug.

Exercise 9.1.2. Recall our list of all B_3-quilts with fewer than 6 seams from Sect. 5.2. Check that the redundancy principle and Thm. 9.1.1 eliminate all of these quilts as possible snug quilts except for the quilts in Figs. 5.2.1 ($M = 1$ only), 5.2.3, and 5.2.5.

Note also that if the modulus of a quilt is very small, the structure of the quilt group is almost completely determined. We have the following theorems to this effect.

Theorem 9.1.3. *The group of a quilt with modulus $M = 1$ is a homomorphic image of the quaternion group of order 8.*

Proof. If $\alpha_1 = \alpha$ and $\beta_1 = \beta$, we have that

$$\alpha\beta\alpha^{-1} = \beta^{-1},$$
$$\beta\alpha\beta^{-1} = \alpha^{-1}. \tag{9.1.1}$$

Coset enumeration then shows that the group with generators α, β and relations (9.1.1) is the quaternion group of order 8. $\square$

Theorem 9.1.4. *If G is the group of a quilt Q with modulus $M = 2$, then G' is cyclic and contained in the center of G. Furthermore, if Q is finite, then G is finite.*

Proof. If $G = \langle \alpha, \beta \rangle$, then $M = 2$ implies that α and β both commute with κ, which is a commutator of α and β. Since G' is generated by the conjugates of κ, the first statement of the theorem follows. As for the second statement, if Q is finite, then for any pair (α, β) of the standard Norton system of G, both α and β have finite order. Following Magnus, Karass, and Solitar [54, p. 23], if β has order n, then by raising both sides of

$$\alpha\beta\alpha^{-1} = \beta\kappa \tag{9.1.2}$$

to the nth power, we see that $1 = \kappa^n$. Since the abelianization of G is obviously finite, it then follows that G is finite. $\square$

Remark 9.1.5. In particular, Thm. 9.1.4 implies that finite abelian quilts have finite groups. In general, however, the group of an abelian quilt may be a nontrivial central extension of an abelian group.

9.2 Results

Having described our methods in the previous section, we now list the small snug quilts, omitting the computational details. Throughout this section, we use s to denote the number of seams in the quilt under discussion, M to denote its modulus, and $G(Q)$ to denote its group. For a given number of seams s, if there is an upper bound noted for M, there may be other snug quilts with higher modulus and the same number of seams; otherwise, the only quilts with s seams are listed here. Also, since all of these quilts happen to have genus 0, Thm. 6.1.10 implies that each quilt is determined by its modular structure, modulus, and patch inflows, so we omit the other arrows of the quilt. In addition, we label each patch with the order of its stack elements. (To avoid confusion with the order of stack elements, we put all inflows in parentheses.)

Notation 9.2.1. The symbol $(\ell, m \mid n, k)$ (Coxeter [21]) denotes the group

$$\langle \alpha, \beta \mid 1 = \alpha^\ell = \beta^m = (\alpha\beta)^n = (\alpha\beta^{-1})^k \rangle. \tag{9.2.1}$$

Note that it is easy to read $(\ell, m \mid n, k)$ relations from the orders of the patches touching the vertices meeting at any edge of a quilt, as shown in Fig. 9.2.1.

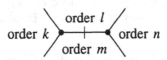

Fig. 9.2.1. Reading $(\ell, m \mid n, k)$ relations from a quilt

s = 1 or 2, M arbitrary. The only snug quilt with 1 or 2 seams is the trivial quilt.

s = 3, M arbitrary. There is an infinite family of quilts with $s = 3$, parameterized by M odd (Fig. 9.2.2). These quilts have

$$G(Q) = \langle \alpha, \beta \mid 1 = \beta^M = \alpha^2, \ \alpha^{-1}\beta\alpha = \beta^{-1} \rangle \cong D_{2M}, \tag{9.2.2}$$

the dihedral group of order $2M$.

s = 4, M arbitrary. There is an infinite family of quilts with $s = 4$ and $M = 2n$, where n is relatively prime to 3, and with $M = 2n$. This family is

Fig. 9.2.2. $M \cdot 3(1_{\frac{1}{2}}, 2)$ **Fig. 9.2.3.** $2n \cdot 4(1_{\frac{2}{3}}, 3)$

shown in Fig. 9.2.3. These quilts have $G(Q) \cong (3, 3 \mid 3, n)$, a group of order $3n^2$. In particular, for $n = 1$, $G(Q) \cong C_3$, and for $n = 2$, $G(Q) \cong A_4$.

s = 5, M arbitrary. There are no snug quilts with 5 seams.

s = 6, M less than 24. For $M < 24$, the snug quilts are part of three infinite families, with two exceptions. The infinite family $M \cdot 6(2, 2_{-i}, 2_{i+1})$ is pictured in Fig. 9.2.4. The family is parameterized by $i > 0$ and M such that M divides $i(i + 1)$. If p and q are the additive orders of i (mod M) and $(i + 1)$ (mod M), respectively, then $G(Q) \cong (2p, 2q \mid 2, 2)$, a group of order $4pq$. Note that if $M = i$, then $G(Q) \cong D_{4i}$.

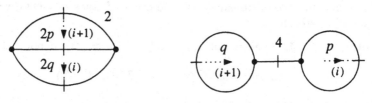

Fig. 9.2.4. $M \cdot 6(2, 2_{-i}, 2_{i+1})$ **Fig. 9.2.5.** $M \cdot 6(1_{-i}, 1_{i+1}, 4)$

The other two infinite families are also of the form $2n \cdot 6(1_{-i}, 1_{i+1}, 4)$ (Fig. 9.2.5), with n odd. (In Fig. 9.2.5, p and q are again the additive orders of i (mod M) and $(i + 1)$ (mod M).) For one family, $i \equiv 0$ (mod 2), $i \equiv \frac{1}{2}$ (mod n), and

$$G(Q) \cong \langle \alpha, \beta \mid 1 = \alpha^4 = \beta^n, \; \alpha^{-1}\beta\alpha = \beta^{-1} \rangle. \tag{9.2.3}$$

For the other family, $i = n$ and $G(Q) \cong (4, 4 \mid 2, n)$, a group of order $4n^2$.

The two small snug quilts not in the above families also have structure $M \cdot 6(1_{-i}, 1_{i+1}, 4)$. For $M = 12$, $i = 3$, the group has shape $2.(A_4 \times A_4) : 4$, and for $M = 15$, $i = 5$, the group has shape $2.(A_5 \times A_5).2$.

s = 7, M less than 35. There is one snug quilt, of shape $7 \cdot 7(3, 4)$, shown in Fig. 9.2.6. We have $G(Q) \cong (3, 3 \mid 4, 4) \cong L_2(7)$ (Coxeter [21]).

s = 8, M arbitrary. There are no snug quilts with 8 seams.

s = 9, M less than 21. With one exception, the snug quilts with 9 seams and $M < 21$ are covers of the quilt in Fig. 9.2.7, whose group is the triangle group $(2\ 3\ 4) \cong S_4$. The covers are part of two infinite families parameterized by n odd, one of which (Fig. 9.2.8) has groups of shape $n^3.S_4$, and the other of which has groups of shape $n.S_4$.

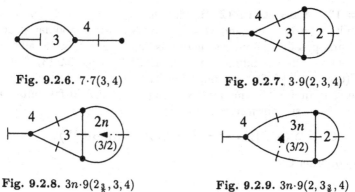

Fig. 9.2.6. 7·7(3, 4)   Fig. 9.2.7. 3·9(2, 3, 4)

Fig. 9.2.8. $3n \cdot 9(2_{\frac{3}{2}}, 3, 4)$ Fig. 9.2.9. $3n \cdot 9(2, 3_{\frac{3}{2}}, 4)$

The aforementioned exception is shown in Fig. 9.2.10. Remarkably, for this quilt,

$$G(Q) \cong \langle \alpha, \beta \mid 1 = \alpha^5 = (\alpha^2 \beta)^3, \; \beta_3 = \beta, \; \alpha\beta = \alpha_6 \rangle \cong U_3(4), \qquad (9.2.4)$$

the simple group of order 62400.

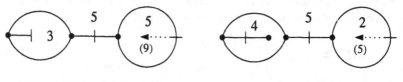

Fig. 9.2.10. $15 \cdot 9(1_9, 3, 5)$ Fig. 9.2.11. $10 \cdot 10(1_5, 4, 5)$

s = 10, M less than 20. There is one snug quilt of shape $10 \cdot 10(1_5, 4, 5)$ (Fig. 9.2.11), with $G(Q) \cong (5, 5 \mid 4, 2) \cong A_6$ (Coxeter [21]).

There are two snug quilts with modular structure $10(2, 3, 5)$. One is the previously discussed $5 \cdot 10(2, 3, 5)$ quilt for A_5 (Fig. 1.3.8), whose group is the triangle group $(2\ 3\ 5) \cong A_5$, and the other is a $10 \cdot 10(2, 3, 5)$ cover of that quilt, shown in Fig. 9.2.12. The latter quilt has a group of shape $2^5.A_5$.

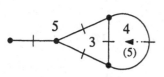

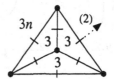

Fig. 9.2.12. $10 \cdot 10(2_5, 3, 5)$ Fig. 9.2.13. $2n \cdot 12(3^3, 3_2)$

s = 11, M arbitrary. There are no snug quilts with 11 seams.

s = 12, M less than 12. Besides the quilt for C_5 (Exmp. 4.4.1), the snug quilts with 12 seams and $M < 12$ have 3 possible modular structures.

The first possible modular structure is $12(3^4)$. For $M < 12$, all such snug quilts fall are part of an infinite family $2n \cdot 12(3^3, 3_2)$ (Fig. 9.2.13). These quilts have $G(Q) \cong (3, 3 \mid 3, 3n)$, a group of order $27n^2$.

The snug quilts with modular structure $12(1, 2, 3, 6)$ fall into 3 infinite families. The first, shown in Fig. 9.2.14, has

$$G(Q) \cong \langle \alpha, \beta \mid 1 = \alpha^2 = \beta^3 = (\alpha\beta)^6 = (\alpha\beta\alpha\beta^2)^n \rangle, \qquad (9.2.5)$$

a group of order $6n^2$.

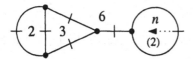
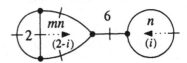

Fig. 9.2.14. $2n \cdot 12(1_2, 2, 3, 6)$ **Fig. 9.2.15.** $2n \cdot 12(1_i, 2, 3_{2-i}, 6)$

The second family, parameterized by n odd, is shown in Fig. 9.2.15, where i is the unique number mod $2n$ such that $i = 0 \pmod 2$ and $i = \frac{1}{2} \pmod n$, and m is 1 if 3 divides n, and 3 otherwise. This family has

$$G(Q) \cong \left\langle \alpha, \beta, \omega \left| \begin{array}{l} 1 = \alpha^6 = \beta^n = \omega^3 = [\beta, \omega], \\ \alpha^{-1}\beta\alpha = \omega\beta^{-1}, \ \alpha^{-1}\omega\alpha = \omega^{-1} \end{array} \right. \right\rangle. \qquad (9.2.6)$$

If n is divisible by 3, then this group has structure $(3 \times n) : 6$; otherwise, (9.2.6) reduces to

$$G(Q) \cong \langle \alpha, \beta \mid 1 = \alpha^6 = \beta^n, \ \alpha^{-1}\beta\alpha = \beta^{-1} \rangle, \qquad (9.2.7)$$

a group of structure $n : 6$.

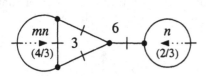
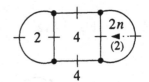

Fig. 9.2.16. $2n \cdot 12(1_{\frac{2}{3}}, 2_{\frac{4}{3}}, 3, 6)$ **Fig. 9.2.17.** $2n \cdot 12(2, 2_2, 4^2)$

The third family (Fig. 9.2.16) has

$$G(Q) \cong \left\langle \alpha, x, y, z \left| \begin{array}{l} 1 = \alpha^6 = x^n = xyz = [x, y], \\ \alpha^{-1}x\alpha = y, \ \alpha^{-1}y\alpha = z, \ \alpha^{-1}z\alpha = x \end{array} \right. \right\rangle, \qquad (9.2.8)$$

a group of shape $n^2:6$, where n is relatively prime to 3. (In Fig. 9.2.16, m is 1 if n is even and 2 if n is odd.)

The other possible modular structure is $12(2^4, 4^2)$. With one exception, these quilts fall into three infinite families. The first family (Fig. 9.2.17) has $G(Q) \cong (4, 4 \mid 2, 2n)$, a group of order $4n^2$.

The second family is parameterized by an integer n, with $M = 2n$. If n is odd (Fig. 9.2.18), then

$$G(Q) \cong \left\langle \alpha, \beta, t \; \middle| \; \begin{matrix} 1 = \alpha^4 = \beta^{2n} = t^2 = [t, \alpha] = [t, \beta], \\ \alpha^{-1}\beta\alpha = t\beta^{-1} \end{matrix} \right\rangle \tag{9.2.9}$$

with structure $(2 \times 2n):4$, and if n is even (Fig. 9.2.19), then

$$G(Q) \cong \left\langle \alpha, \beta \mid 1 = \alpha^4 = \beta^{4n}, \; \alpha^{-1}\beta\alpha = \beta^{-1}, \; \alpha^2 = \beta^{2n} \right\rangle, \tag{9.2.10}$$

with structure $4n.2$.

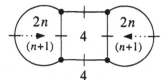

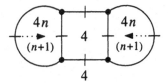

Fig. 9.2.18. $2n \cdot 12(2_{n+1}^2, 4^2)$, n odd   **Fig. 9.2.19.** $2n \cdot 12(2_{n+1}^2, 4^2)$, n even

The last family (Fig. 9.2.20) is parameterized by an integer n, with

$$G(Q) = \left\langle \alpha, \beta \mid 1 = \alpha^4 = (\alpha\beta)^2 = \beta^{4n} = [\beta, \alpha^2] \right\rangle, \tag{9.2.11}$$

a group of shape $2.D_{8n}$.

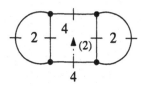

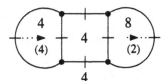

Fig. 9.2.20. $2n \cdot 12(2^2, 4, 4_2)$ **Fig. 9.2.21.** $8 \cdot 12(2^2, 4^2)$

The remaining snug quilt with $s = 12$ is shown in Fig. 9.2.21. Its group has structure $4^2.(2 \times 4)$.

10. Monodromy systems

This chapter describes *mondromy systems*, or *M(n)-systems*, which are a version of Norton systems arising from monodromy representations of $\mathbf{B}_n$. We show (Thms. 10.2.4 and 10.2.5) that in some sense, all Norton systems are equivalent to a special type of *M(3)-system*, called an *involution M(3)-system*.

All braid-related notation in this chapter is from Sect. 2.1.

10.1 Monodromy representations of $\mathbf{B}_n$

Throughout this section, we fix a group G. We have the following theorem.

Theorem 10.1.1. *The map sending σ_i to the transformation*

$$(\ldots, g_i, g_{i+1}, \ldots) \mapsto (\ldots, g_i g_{i+1} g_i^{-1}, g_i, \ldots) \qquad (10.1.1)$$

determines a right action of $\mathbf{B}_n$ on the set of n-tuples of elements of G.

Proof. A calculation shows that (10.1.1) respects the defining relations in presentation (2.4.1). □

The action of $\mathbf{B}_n$ induced by (10.1.1) is called the *monodromy action*. (Motivation for (10.1.1), and for the use of the term "monodromy", can be found in Birman [6, §1.4].)

Definition 10.1.2. An orbit of the monodromony action of $\mathbf{B}_n$ on n-tuples of elements of G is called a *monodromy system*, or *M(n)-system*, in G.

Remark 10.1.3. We remark that the monodromy action and monodromy systems have become crucial to the study of the Inverse Galois Problem. See Völklein [86, Chap. 9] for an introduction.

10.2 Involution M(3)-systems

Henceforth, we only consider the following type of M(n)-system.

Definition 10.2.1. Let G be a group. An *involution M(n)-system* in G is the M(n)-system of an n-tuple of involutions of G. Note that because monodromy merely permutes the elements of an n-tuple, up to conjugacy, all of the n-tuples of an involution M(n)-system are n-tuples of involutions.

It is often easier to work with a reduced version of involution M(n)-systems. Let G be a group, and let r be the function from n-tuples in G to $(n-1)$-tuples in G defined by

$$r(a_1, a_2, \ldots, a_n) = (a_1 a_2, a_2 a_3, \ldots, a_{n-1} a_n). \tag{10.2.1}$$

If we let $\alpha_i = a_i a_{i+1}$, we arrive naturally at the following.

Definition 10.2.2. The *reduced monodromy action* of $\mathbf{B}_n$ on $(n-1)$-tuples of elements of G is defined by

$$(\alpha_1, \alpha_2, \ldots)\sigma_1 = (\alpha_1, \alpha_1 \alpha_2, \ldots), \tag{10.2.2}$$

$$(\ldots, \alpha_{i-1}, \alpha_i, \alpha_{i+1}, \ldots)\sigma_i = (\ldots, \alpha_{i-1}\alpha_i^{-1}, \alpha_i, \alpha_i \alpha_{i+1}, \ldots), \tag{10.2.3}$$

$$(\ldots, \alpha_{n-2}, \alpha_{n-1})\sigma_{n-1} = (\ldots, \alpha_{n-2}\alpha_{n-1}^{-1}, \alpha_{n-1}). \tag{10.2.4}$$

for $1 < i < n-1$. An orbit of this action is called a *reduced monodromy system*.

It is easily verified that (10.2.2)–(10.2.4) are simply the result of applying the reduction r to the monodromy action on n-tuples of involutions, so reduced monodromy is a well-defined action of $\mathbf{B}_n$. In particular, for $n = 3$, we get $(\alpha, \beta)\sigma_1 = (\alpha, \alpha\beta)$ and $(\alpha, \beta)\sigma_2 = (\alpha\beta^{-1}, \beta)$, which is precisely Norton's action (see Sect. 4.2). In other words, a reduced M(3)-system is a Norton system. Note also that stacking (Defn. 4.2.10) in a reduced M(3)-system is just conjugation, since $Z = (LR^{-1}L)^2 = (\sigma_1 \sigma_2 \sigma_1)^2$ and

$$(a, b, c)(\sigma_1 \sigma_2 \sigma_1)^2 = (a^{cba}, b^{cba}, c^{cba}). \tag{10.2.5}$$

Since the product cba is invariant under the action of $\mathbf{B}_3$, we define cba to be the *monodromy stacking element* of the M(3)-system of (a, b, c). Note that if $N = N(ab, bc)$ is the reduction of the M(3)-system of (a, b, c), then the stacking element (Thm. 4.2.9) of N is precisely $(bc)^{-1}(ab)(bc)(ab)^{-1} = (cba)^2$, and the modulus of N is the smallest positive integer M such that $(cba)^M$ centralizes both ab and bc.

Now, since any reduced M(3)-system is a Norton system, we can ask the question: when does a Norton system arise in this manner? We have the following answers.

Proposition 10.2.3. *A Norton system N for a group G is the reduction of an involution M(3)-system in some group G_0 containing G if and only if for some pair $(\alpha, \beta) \in N$, there is an involution $b \in G$ such that $b\alpha b = \alpha^{-1}$ and $b\beta b = \beta^{-1}$.*

Proof. If N is a reduction of an involution M(3)-system M, then any pair in N has the form (ab, bc) for some $(a, b, c) \in M$, and conjugation by b clearly inverts ab and bc. Conversely, if there is an inverting involution b for some pair $(\alpha, \beta) \in N$, then the M(3)-system of $(\alpha b, b, b\beta)$ is an involution M(3)-system that reduces to N. □

Theorem 10.2.4. *Let N be a Norton system for a group G. There exists a group G_1 containing G and a Norton system N_0 for some $G_0 \leq G_1$ such that N_0 is isomorphic to N, N_0 is the reduction of an M(3)-system in G_1, and G_0 surjects onto G. In particular, if G is the group of a snug quilt Q, we can choose $G_1 \cong G{:}2$.*

In other words, any Norton system comes from an M(3)-system, if we allow ourselves to "enlarge" the ambient group.

Proof. Recall (Thm. 4.5.1) that

$$\mathrm{Stab}_{\mathbf{B}_3}(\alpha, \beta) = \mathrm{Stab}_{\mathbf{B}_3}(\alpha^{-1}, \beta^{-1}). \tag{10.2.6}$$

Let G_1 be the wreath product $G \wr C_2$, with wreathing involution b, and let $N_0 = N((\alpha, \alpha^{-1}), (\beta, \beta^{-1}))$. From (10.2.6), it follows that N_0 is isomorphic to N, and since b inverts (α, α^{-1}) and (β, β^{-1}), N_0 is a reduced M(3)-system in G_1 (Prop. 10.2.3). Therefore, if we let $G_0 = \langle (\alpha, \alpha^{-1}), (\beta, \beta^{-1}) \rangle$, then G_0 clearly surjects onto $\langle \alpha, \beta \rangle$, and the first statement of the theorem follows. As for the second statement, in that case, by the "universal" property of the group of a snug quilt (Thm. 8.1.4), G must surject onto G_0, and is therefore isomorphic to G_0. Since we can choose G_1 to be the subgroup of $G \wr C_2$ generated by G_0 and b, the rest of the theorem follows. □

For instance, if G is the cyclic group of order n, written multiplicatively with generator x, and N is the Norton system of $(x, 1)$, then we can take G_0 to be $D_{2n} \cong \langle x, b \,|\, 1 = x^n = b^2,\ b^{-1}xb = x^{-1} \rangle$. In fact, N is the reduction of the M(3)-system of (xb, b, b).

Next, we turn to the question of whether structural information is lost when we reduce an involution M(n)-system. More precisely, does

$$\mathrm{Stab}_{\mathbf{B}_n}(a_1, a_2, \ldots, a_n) = \mathrm{Stab}_{\mathbf{B}_n}(a_1 a_2, a_2 a_3, \ldots, a_{n-1} a_n)? \tag{10.2.7}$$

Curiously, the answer to (10.2.7) is yes when n is odd, and no when n is even. We begin with the case $n = 3$.

Theorem 10.2.5. *For involutions $a, b, c \in G$,*

$$\mathrm{Stab}_{\mathbf{B}_3}(a, b, c) = \mathrm{Stab}_{\mathbf{B}_3}(ab, bc). \tag{10.2.8}$$

Proof. The proof is essentially a straightforward calculation. Let $A = ab$ and $B = bc$. We first note that since the braid action commutes with reduction,

$\mathrm{Stab}_{\mathbf{B}_3}(a,b,c) \subset \mathrm{Stab}_{\mathbf{B}_3}(A,B)$. It therefore suffices to show that for any $\pi \in \mathbf{B}_3$, $(A,B)\pi = (A,B)$ implies $(a,b,c)\pi = (a,b,c)$.

Let π be an arbitrary element of $\mathrm{Stab}_{\mathbf{B}_3}(A,B)$. Since $\mathbf{B}_3 = \langle L, R \rangle$, we know that

$$\pi = L^{\ell_1} R^{r_1} L^{\ell_2} R^{r_2} \cdots L^{\ell_N} R^{r_N} \tag{10.2.9}$$

for some positive integer N and some $\ell_i, r_i \in \mathbf{Z}$. Define

$$A_0 = A, \qquad B_0 = B, \tag{10.2.10}$$
$$B_{i+1} = A_i^{\ell_{i+1}} B_i, \quad A_{i+1} = A_i B_{i+1}^{r_{i+1}},$$

for $0 \le i < N$. Applying L^{ℓ_i} and then R^{r_i}, for $1 \le i \le N$, we get

$$(A,B)\pi = (AB_1^{r_1} B_2^{r_2} \cdots B_N^{r_N}, A_{N-1}^{\ell_N} A_{N-2}^{\ell_{N-1}} \cdots A_0^{\ell_1} B), \tag{10.2.11}$$

which means that

$$1 = B_1^{r_1} B_2^{r_2} \cdots B_N^{r_N} = A_{N-1}^{\ell_N} A_{N-2}^{\ell_{N-1}} \cdots A_0^{\ell_1}. \tag{10.2.12}$$

(We will see later that this is where we use the fact that n is odd.)

We now come to the key point, which is that for all integers ℓ_1 and r_1,

$$(a,b,c)L^{\ell_1}R^{r_1} = (aA_0^{-\ell_1}, A_0^{\ell_1} b, c)R^{r_1} \tag{10.2.13}$$
$$= (aA_0^{-\ell_1}, A_0^{\ell_1} bB_1^{r_1}, B_1^{-r_1} c).$$

In other words, when viewed appropriately, the monodromy action on involutions is really just either left or right multiplication. We therefore define

$$a_0 = a, \qquad\qquad b_0 = b, \qquad\qquad c_0 = c, \tag{10.2.14}$$
$$a_{i+1} = a_i A_i^{-\ell_{i+1}}, \qquad b_{i+1} = A_i^{\ell_{i+1}} b_i B_{i+1}^{r_{i+1}}, \qquad c_{i+1} = B_{i+1}^{-r_{i+1}} c_i,$$

where A_i and B_i are as before. Applying (10.2.13), we get

$$(a,b,c)\pi = (a_N, b_N, c_N)$$
$$= (aA_0^{-\ell_1} A_1^{-\ell_2} \cdots A_{N-1}^{-\ell_N},$$
$$A_{N-1}^{\ell_N} A_{N-2}^{\ell_{N-1}} \cdots A_0^{\ell_1} bB_1^{r_1} B_2^{r_2} \cdots B_N^{r_N}, \tag{10.2.15}$$
$$B_N^{-r_N} B_{N-1}^{-r_{N-1}} \cdots B_1^{-r_1} c)$$
$$= (a,b,c). \quad \square$$

We will subsequently treat M(3)-systems and Norton systems as structurally interchangable. In particular,

Corollary 10.2.6. *An M(3)-system has the same quilt as its reduction.*

For n odd and greater than 3, the proof of the statement analogous to Thm. 10.2.5 is quite similar, but notationally awkward, so we will omit it. To give an idea of what the proof looks like, we mention that for $n = 5$, the step analogous to (10.2.13) looks something like

$$(a, b, c, d, e) \mapsto (aA^{-1}, AbB^{-1}, BcC^{-1}, CdD^{-1}, De), \qquad (10.2.16)$$

and the step analogous to (10.2.12) looks something like

$$\begin{aligned} (\alpha, \beta, \gamma, \delta) = (\alpha B^{-1}, A\beta C^{-1}, B\gamma D^{-1}, C\delta) \quad \text{implies} \\ 1 = B = D \text{ and } 1 = C = A. \end{aligned} \qquad (10.2.17)$$

Note that for the cancellation in (10.2.17) to work completely, n must be odd.

In contrast, when n is even, it is easy to see that the analogy of Thm. 10.2.5 does not hold. Let G be the infinite dihedral group

$$\langle r, s \mid 1 = r^2 = s^2 \rangle. \qquad (10.2.18)$$

If n is even, then

$$(r, s, r, s, \ldots, r, s)\sigma_1\sigma_3 \cdots \sigma_{n-1} = (rsr, r, rsr, r, \ldots, rsr, r). \qquad (10.2.19)$$

However, reducing (10.2.19) on both sides, we get

$$(rs, sr, \ldots, rs)\sigma_1\sigma_3 \cdots \sigma_{n-1} = (rs, sr, \ldots, rs). \qquad (10.2.20)$$

Iterating the action in (10.2.19), we see that the involution stabilizer has infinite index in the reduced stabilizer. In fact, if we take G to be an arbitrarily large dihedral group instead of an infinite one, we see that even if G is finite, the involution stabilizer can have arbitrarily large index in the reduced stabilizer.

11. Quilts for groups involved in the Monster

In this chapter, we examine some examples of quilts for groups involved in
M, the Fischer-Griess Monster group. Specifically, we consider groups that
are the non-abelian composition factors of centralizers of elements of M, and
we exhibit some (or in some cases, all) of their quilts that arise from a 6-
transposition construction due to Norton (Sect. 11.1).

11.1 Quilts from n-transpositions

The original motivation for studying quilts came from the following construction of Norton [62]. Suppose a group G has a conjugacy class C of involutions
such that the product of any two elements of C has order $\leq n$. The elements
of C are then known as n-transpositions. For instance, it can be shown, using
character theory, that conjugacy class $2A$ (in ATLAS [16] notation) of M is
a class of 6-transpositions.

We define an n-transposition quilt to be the quilt of an M(3)-system of n-
transpositions. The reason n-transposition quilts are particularly interesting
is that their patches are all of order $\leq n$. Therefore, from curvature results
(Cor. 6.2.3) or the Riemann-Hurwitz formula, a 6-transposition quilt has
genus 0 or 1; in fact, a 6-transposition has genus 1 if and only if it is collapse-
free and all of its patches have order 6. This fact is an interesting "genus zero
property" of 6-transposition groups, and since the Monster is thought to be
(in some sense) a maximal 6-transposition group, one might hope that this
might be related to the *genus zero phenomenon* of Monstrous Moonshine (see
Sect. 1.1).

In this chapter, we examine 6-transposition quilts for certain quotients of
subgroups of M. Most of the groups we consider are of the form F_m, where m
is a conjugacy class of the Monster, and F_m is defined to be the non-abelian
composition factor of the centralizer of an element of m. For the convenience
of the reader, in Table 11.1.1, we list the names we have used (following
Conway and Norton [18]) for conjugacy classes of the Monster, compared
with their ATLAS names.

Table 11.1.2 lists (in ATLAS notation) the classes of 6-transpositions we
have examined in each group we consider. The notation $2AB$ means that the
conjugacy classes $2A$ and $2B$ are both part of a single 6-transposition class

Table 11.1.1. Moonshine vs. ATLAS names in the Monster

Class	ATLAS	Class	ATLAS	Class	ATLAS
$2-$	$2B$	$6+$	$6A$	$8|2-$	$8D$
$3-$	$3B$	$6+3$	$6C$	$10+$	$10A$
$4+$	$4A$	$6-$	$6E$	$10-$	$10E$
$4-$	$4C$	$6|3$	$6F$	$11+$	$11A$
$4|2-$	$4D$	$7-$	$7B$	$13+$	$13A$
$5-$	$5B$	$8+$	$8A$	$17+$	$17A$

inside the Monster, and the notation $2A, 2B$ means that classes $2A$ and $2B$ are 6-transposition classes separately, but not together.

Table 11.1.2. Classes of 6-transpositions

In **M**	Group	Class(es)	In **M**	Group	Class(es)
F_{2-}	Co_1	$2A$		A_{12}	$2AB$
	Co_2	$2B$	$F_{6|3}$	A_9	$2A$
F_{3-}	Suz	$2A$		A_8	$2AB$
F_{4+}	Co_3	$2A$	F_{7-}	A_7	$2A$
F_{4-}	$W(E_7)'$	$2ABC$	F_{8+}	$U_3(3)$	$2A$
$F_{4|2-}$	$G_2(4)$	$2A$	$F_{8|2-}$	A_6	$2A$
	$U_3(4)$	$2A$	F_{10+}	HS	$2A$
F_{5-}	J_2	$2A$	F_{10-}	A_5	$2A$
F_{6+}	Fi_{22}	$2B$	F_{11+}	M_{12}	$2A, 2B$
	$U_6(2)$	$2B$	F_{13+}	$L_3(3)$	$2A$
F_{6+3}	$U_4(3)$	$2A$	F_{17+}	$L_3(2)$	$2A$
F_{6-}	$U_4(2)$	$2AB$			

Most of the groups in this chapter were constructed explicitly from their descriptions in the ATLAS and Conway and Sloane [20, Chap. 10–11], using the computational group theory system GAP [77]. The groups J_2, $G_2(4)$, and Suz were constructed from their descriptions in Hall and Wales [38] and Suzuki [83], using GAP, L.H. Soicher's program GRAPE [81], and B.D. McKay's program *nauty* [58].

In Sects. 11.2–11.4, we present only a sampling of the 6-transposition quilts for the groups we examine, not a complete enumeration. We indicate this situation by saying something like "*Some* of the quilts ...." On the other hand, in Sects. 11.5 and 11.6, we enumerate *all* of the 6-transposition quilts for the groups in question.

Conventions for drawing quilts. Throughout this chapter, we continue our practice of only indicating non-zero patch inflow arrows on a quilt of genus 0. (Recall from Cor. 6.1.11 that the modulus, modular structure and patch inflows of a genus 0 quilt identify it uniquely.) Furthermore, we omit the dashes on uncollapsed edges. A number in parentheses indicates inflow, and any other number inside a patch indicates the order of a stack element for that patch. We also use the identification notation described in Sect. 5.1.

11.2 Conway's groups Co$_n$

Some of the 6-transposition quilts for Co$_1$ are shown in Figs. 11.2.1–11.2.10. We call particular attention to the $30 \cdot 180(5^{12}, 6^{20})$ quilt in Fig. 11.2.9, which has the same proportion of pentagons and hexagons as a soccer ball, or outside of the United States, a football. It is the possibility of quilts such as this one that led to quilts' original name of *footballs*. (Recently, Norton [66, 68] has proposed calling 6-transposition quilts *nets* and genus zero 6-transposition quilts *netballs*; see Sect. 13.1 for details.)

The group Co$_2$ is not of the form F_m, but we include it for completeness. Some of its 6-transposition quilts are shown in Figs. 11.2.11–11.2.15. Finally, Figs. 11.2.16–11.2.25 show some 6-transposition quilts for Co$_3$.

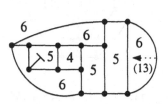

Fig. 11.2.1. Co$_1$:
$39 \cdot 39(2_{13}, 4, 5^3, 6^3)$

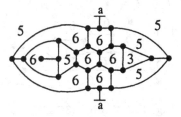

Fig. 11.2.2. Co$_1$: $35 \cdot 70(3, 5^5, 6^7)$

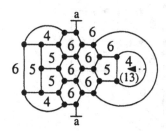

Fig. 11.2.3. Co$_1$:
$26 \cdot 78(2_{13}, 4^2, 5^4, 6^8)$

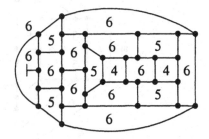

Fig. 11.2.4. Co$_1$: $33 \cdot 99(4^2, 5^5, 6^{11})$

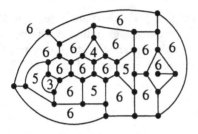

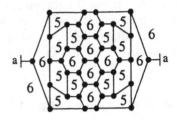

Fig. 11.2.5. Co_1:
$28 \cdot 112(3, 4, 5^3, 6^{15})$

Fig. 11.2.6. Co_1: $20 \cdot 120(5^{12}, 6^{10})$

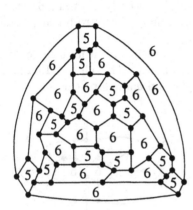

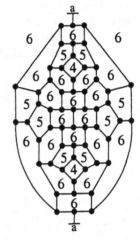

Fig. 11.2.7. Co_1: $13 \cdot 156(5^{12}, 6^{16})$

Fig. 11.2.8. Co_1: $14 \cdot 168(4^2, 5^8, 6^{20})$

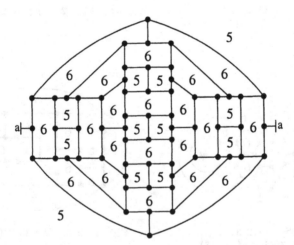

Fig. 11.2.9. Co_1: $30 \cdot 180(5^{12}, 6^{20})$

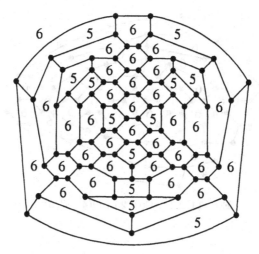

Fig. 11.2.10. Co$_1$: $21 \cdot 252(5^{12}, 6^{32})$

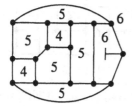

Fig. 11.2.11. Co$_2$: $15 \cdot 45(4^2, 5^5, 6^2)$

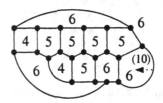

Fig. 11.2.12. Co$_2$:
$20 \cdot 60(3_{10}, 4^2, 5^5, 6^4)$

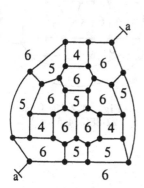

Fig. 11.2.13. Co$_2$: $16 \cdot 96(4^3, 5^6, 6^9)$

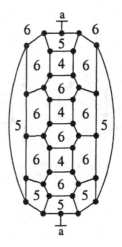

Fig. 11.2.14. Co$_2$:
$18 \cdot 108(4^3, 5^6, 6^{11})$

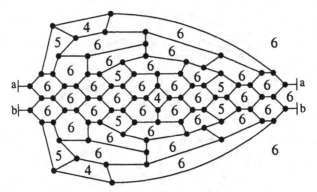

Fig. 11.2.15. Co_2: $20 \cdot 240(4^3, 5^6, 6^{33})$

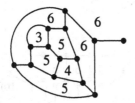

Fig. 11.2.16. Co_3: $20 \cdot 40(3, 4, 5^3, 6^3)$

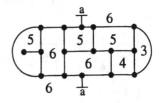

Fig. 11.2.17. Co_3: $23 \cdot 46(3, 4, 5^3, 6^4)$

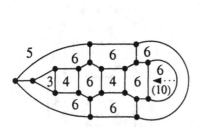

Fig. 11.2.18. Co_3:
$30 \cdot 60(2_{10}, 3, 4^2, 5, 6^7)$

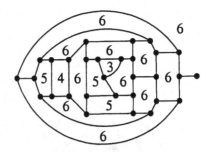

Fig. 11.2.19. Co_3:
$22 \cdot 88(3, 4, 5^3, 6^{11})$

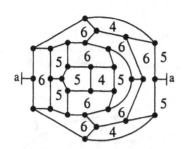

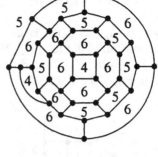

Fig. 11.2.20. Co$_3$: $15 \cdot 90(4^3, 5^6, 6^8)$

Fig. 11.2.21. Co$_3$:
$18 \cdot 108(4^2, 5^8, 6^{10})$

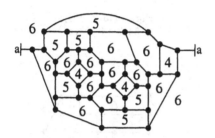

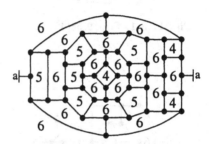

Fig. 11.2.22. Co$_3$:
$20 \cdot 120(4^3, 5^6, 6^{13})$

Fig. 11.2.23. Co$_3$:
$24 \cdot 144(4^3, 5^6, 6^{17})$

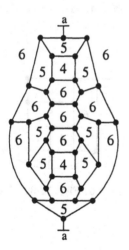

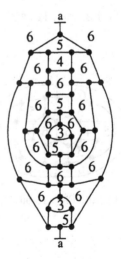

Fig. 11.2.24. Co$_3$:
$18 \cdot 108(4^2, 5^8, 6^{10})$

Fig. 11.2.25. Co$_3$:
$21 \cdot 126(3^2, 4, 5^4, 6^{16})$

11.3 Fischer's groups Fi_{21} and Fi_{22}

The group $\mathrm{Fi}_{21} \cong U_6(2)$ is not an F_m, but we include some examples of 6-transposition quilts for Fi_{21} (Figs. 11.3.1–11.3.5) anyway, since Fi_{21} is involved in several F_m. In addition, some 6-transposition quilts for Fi_{22} are shown in Figs. 11.3.6–11.3.15.

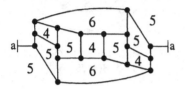

Fig. 11.3.1. Fi_{21}: $9 \cdot 54(4^3, 5^6, 6^2)$

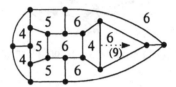

Fig. 11.3.2. Fi_{21}: $18 \cdot 54(3_9, 4^3, 5^3, 6^4)$

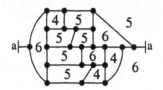

Fig. 11.3.3. Fi_{21}: $11 \cdot 66(4^3, 5^6, 6^4)$

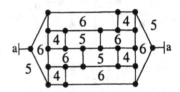

Fig. 11.3.4. Fi_{21}: $12 \cdot 72(4^4, 5^4, 6^6)$

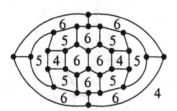

Fig. 11.3.5. Fi_{21}: $15 \cdot 90(4^3, 5^6, 6^8)$

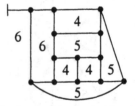

Fig. 11.3.6. Fi_{22}: $13 \cdot 39(4^3, 5^3, 6^2)$

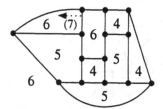

Fig. 11.3.7. Fi$_{22}$:
$14 \cdot 42(3_7, 4^3, 5^3, 6^2)$

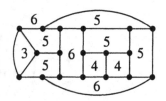

Fig. 11.3.8. Fi$_{22}$: $9 \cdot 54(3, 4^2, 5^5, 6^3)$

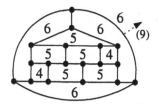

Fig. 11.3.9. Fi$_{22}$:
$18 \cdot 54(3_9, 4^2, 5^5, 6^3)$

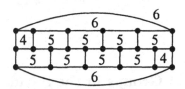

Fig. 11.3.10. Fi$_{22}$: $11 \cdot 66(4^2, 5^8, 6^3)$

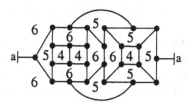

Fig. 11.3.11. Fi$_{22}$: $13 \cdot 78(4^3, 5^6, 6^6)$

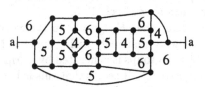

Fig. 11.3.12. Fi$_{22}$: $13 \cdot 78(4^3, 5^6, 6^6)$

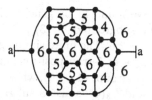

Fig. 11.3.13. Fi$_{22}$: $15 \cdot 90(4^2, 5^8, 6^7)$

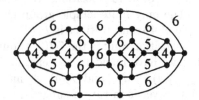

Fig. 11.3.14. Fi$_{22}$:
$18 \cdot 108(4^4, 5^4, 6^{12})$

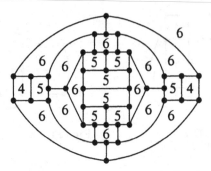

Fig. 11.3.15. Fi$_{22}$: 21·126($4^2, 5^8, 6^{13}$)

11.4 Groups in the Suzuki chain

Some 6-transposition quilts for Suz are shown in Figs. 11.4.1–11.4.10. Note that the $15 \cdot 90(4^3, 5^6, 6^8)$ 6-transposition quilt for Fi$_{21}$ (Fig. 11.3.5) also happens to be a 6-transposition quilt for Suz. For a discussion of this phenomenon, see Sect. 12.3.

The other F_m groups in the Suzuki chain, $U_3(3)$, J$_2$, and $G_2(4)$, behave somewhat unusually with respect to Monstrous 6-transpositions, in that none of these groups is generated by a triple of 6-transpositions, as verified by computer search. It seems that this is because the Monstrous 6-transpositions become too "short" when working in these groups. Specifically, upon restricting to $U_3(3)$, J$_2$, or $G_2(4)$, the Monstrous 6-transposition class becomes a class of 4-, 5-, or 5-transpositions, respectively.

The only finite simple group that we have found to be generated by a triple of 4-transpositions is $L_3(2)$ (a maximal subgroup of $U_3(3)$), and its 4-transposition quilt is the $7 \cdot 7(3, 4)$ quilt in Fig. 9.2.6. The finite simple groups that we have found to be generated by a triple of 5-transpositions are A_5, A_6, $L_3(4) \cong M_{21}$, and $U_3(4)$. A_5 and A_6 are subgroups of $L_3(4)$, whose Schur triple cover is in turn a subgroup of $G_2(4)$, and $U_3(4)$ is also a subgroup of $G_2(4)$. The 5-transposition quilts of A_5 and A_6 are discussed in Sect. 11.5, and the 5-transposition quilts of $L_3(4)$ are discussed in [44]. As for $U_3(4)$, it has three 5-transposition quilts: the $15 \cdot 9(1_9, 3, 5)$ quilt in Fig. 9.2.10, the $5 \cdot 60(5^{12})$ dodecahedral quilt without patch inflows (see Sect. 12.4), and the $13 \cdot 13(3, 5^2)$ quilt in Fig. 11.4.11.

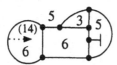

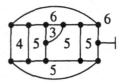

Fig. 11.4.1. Suz: $21 \cdot 21(2_{14}, 3, 5^2, 6)$

Fig. 11.4.2. Suz: $13 \cdot 39(3, 4, 5^4, 6^2)$

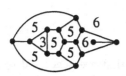

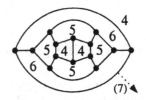

Fig. 11.4.3. Suz: $20 \cdot 40(3, 5^5, 6^2)$

Fig. 11.4.4. Suz: $14 \cdot 42(2_7, 4^2, 5^4, 6^2)$

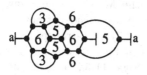

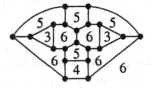

Fig. 11.4.5. Suz: $15 \cdot 45(3^2, 5^3, 6^4)$

Fig. 11.4.6. Suz: $10 \cdot 60(3^2, 4, 5^4, 6^5)$

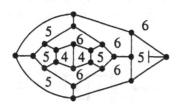

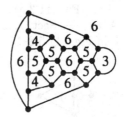

Fig. 11.4.7. Suz: $21 \cdot 63(4^2, 5^5, 6^5)$

Fig. 11.4.8. Suz: $11 \cdot 66(3, 4^2, 5^5, 6^5)$

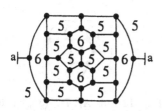

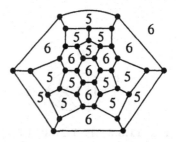

Fig. 11.4.9. Suz: $7 \cdot 84(5^{12}, 6^4)$

Fig. 11.4.10. Suz: $9 \cdot 108(5^{12}, 6^8)$

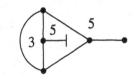

Fig. 11.4.11. $U_3(4)$: $13 \cdot 13(3, 5^2)$

11.5 Alternating groups

The simple alternating groups involved in M that are generated by a triple of 6-transpositions are A_5, A_6, A_7, A_8, and A_{12}. Note that we consider A_8 instead of A_9 because A_9 is not generated by a triple of 6-transpositions. Similarly, although A_{12} is not a composition factor of a Monster centralizer, it is the largest alternating group involved in the Monster, and plays an important role in the Y_{555} construction of the Monster (see Conway and Pritchard [19], Ivanov [46] and Norton [63]), so we consider it here.

Both of the quilts for A_5 (Figs. 1.3.8 and 4.4.7) are 6-transposition quilts. The 6-transposition quilts for A_6 are shown in Figs. 11.5.1 and 11.5.2, the 6-transposition quilts for A_7 are shown in Figs. 11.5.3–11.5.4, and the 6-transposition quilts for A_8 are shown in Figs. 11.5.5–11.5.7. As for A_{12}, we have already seen two of its 6-transposition quilts as the $15 \cdot 90(4^3, 5^6, 6^8)$ quilt for Fi$_{21}$ (Fig. 11.3.5) and the $11 \cdot 66(3, 4^2, 5^5, 6^5)$ quilt for Suz (Fig. 11.4.8). (Again, see Sect. 12.3.) The remaining 6-transposition quilts for A_{12} are shown in Figs. 11.5.8–11.5.23.

Fig. 11.5.1. A_6: $5 \cdot 15(3^2, 4, 5)$

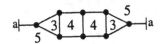

Fig. 11.5.2. A_6: $4 \cdot 24(3^2, 4^2, 5^2)$

Fig. 11.5.3. A_7: $7 \cdot 21(3^2, 4, 5, 6)$

Fig. 11.5.4. A_7: $7 \cdot 21(3^2, 4, 5, 6)$

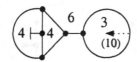

Fig. 11.5.5. A_8: $15 \cdot 15(1_{10}, 4^2, 6)$

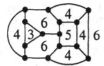

Fig. 11.5.6. A_8: $7 \cdot 42(3, 4^4, 5, 6^3)$

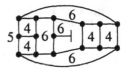

Fig. 11.5.7. A_8: $15 \cdot 45(4^4, 5, 6^4)$

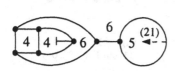

Fig. 11.5.8. A_{12}: $35 \cdot 21(1_{21}, 4^2, 6^2)$

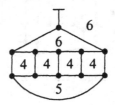

Fig. 11.5.9. A_{12}: $11 \cdot 33(4^4, 5, 6^2)$

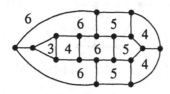

Fig. 11.5.10. A_{12}: $9 \cdot 54(3, 4^3, 5^3, 6^4)$

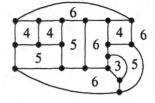

Fig. 11.5.11. A_{12}: $9 \cdot 54(3, 4^3, 5^3, 6^4)$

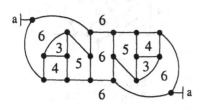

Fig. 11.5.12. A_{12}:
$9 \cdot 54(3^2, 4^2, 5^2, 6^5)$

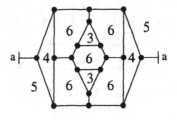

Fig. 11.5.13. A_{12}:
$9 \cdot 54(3^2, 4^2, 5^2, 6^5)$

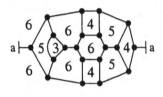

Fig. 11.5.14. A_{12}: $10 \cdot 60(3, 4^3, 5^3, 6^5)$

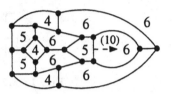

Fig. 11.5.15. A_{12}: $20 \cdot 60(3_{10}, 4^3, 5^3, 6^5)$

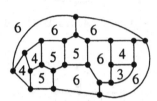

Fig. 11.5.16. A_{12}: $11 \cdot 66(3, 4^3, 5^3, 6^6)$

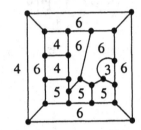

Fig. 11.5.17. A_{12}: $11 \cdot 66(3, 4^3, 5^3, 6^6)$

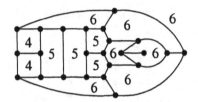

Fig. 11.5.18. A_{12}: $35 \cdot 70(4^2, 5^4, 6^7)$

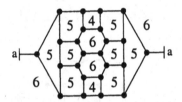

Fig. 11.5.19. A_{12}: $12 \cdot 72(4^2, 5^8, 6^4)$

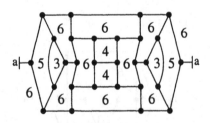

Fig. 11.5.20. A_{12}: $14 \cdot 84(3^2, 4^2, 5^2, 6^{10})$

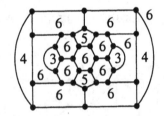

Fig. 11.5.21. A_{12}: $8 \cdot 96(3^2, 4^2, 5^2, 6^{12})$

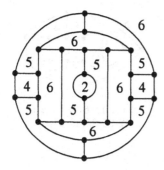

Fig. 11.5.22. A_{12}:
$12 \cdot 72(2, 4^2, 5^4, 6^7)$

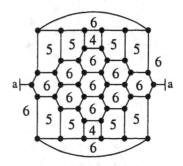

Fig. 11.5.23. A_{12}:
$21 \cdot 126(4^2, 5^8, 6^{13})$

11.6 Some other F_m

This section describes 6-transposition quilts for some other F_m that do not fall in the above categories.

It turns out that the group $F_{4-} \cong W(E_7)'$ is not generated by a triple of 6-transpositions, but its analogue $F_{6-} \cong U_4(2) \cong W(E_6)'$ is, so we exhibit the 6-transposition quilts for $U_4(2)$ in Figs. 11.6.1–11.6.6. Next, the 6-transposition quilts for $F_{10+} \cong HS$ are shown in Figs. 11.6.7–11.6.14. Finally, we have $F_{13+} \cong L_3(3)$, whose 6-transposition quilts are shown in Figs. 11.6.15 and 11.6.16.

The Mathieu group M_{12} is involved in the Monster as F_{11+}. Its 6-transposition quilts, along with the 6-transposition quilts of the other Mathieu groups, are enumerated in [44].

Fig. 11.6.1. $U_4(2)$: $9 \cdot 27(3, 4^2, 5^2, 6)$

Fig. 11.6.2. $U_4(2)$: $9 \cdot 27(3, 4^2, 5^2, 6)$

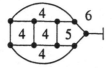

Fig. 11.6.3. $U_4(2)$: $9 \cdot 27(4^4, 5, 6)$

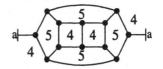

Fig. 11.6.4. $U_4(2)$: $6 \cdot 36(4^4, 5^4)$

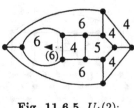

Fig. 11.6.5. $U_4(2)$:
$12 \cdot 36(3_6, 4^4, 5, 6^2)$

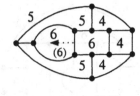

Fig. 11.6.6. $U_4(2)$:
$12 \cdot 36(3_6, 4^3, 5^3, 6)$

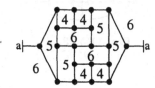

Fig. 11.6.7. HS: $10 \cdot 60(4^4, 5^4, 6^4)$

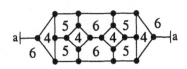

Fig. 11.6.8. HS: $10 \cdot 60(4^4, 5^4, 6^4)$

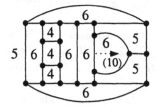

Fig. 11.6.9. HS:
$20 \cdot 60(3_{10}, 4^3, 5^3, 6^5)$

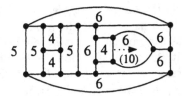

Fig. 11.6.10. HS:
$20 \cdot 60(3_{10}, 4^3, 5^3, 6^5)$

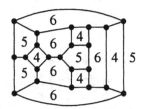

Fig. 11.6.11. HS: $11 \cdot 66(4^4, 5^4, 6^5)$

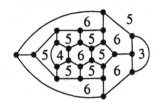

Fig. 11.6.12. HS: $12 \cdot 72(3, 4, 5^7, 6^5)$

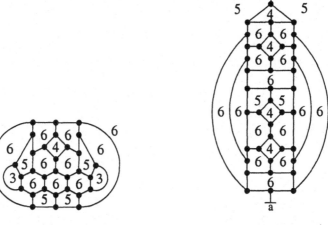

Fig. 11.6.13. HS:
$15 \cdot 90(3^2, 4, 5^4, 6^{10})$

Fig. 11.6.14. HS: $10 \cdot 120(4^4, 5^4, 6^{14})$

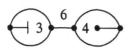

Fig. 11.6.15. $L_3(3)$: $13 \cdot 13(3, 4, 6)$

Fig. 11.6.16. $L_3(3)$: $13 \cdot 13(3, 4, 6)$

11.7 Groups of 6-transposition quilts

In closing, we remark that coset enumeration shows that the 6-transposition quilts for A_5, $L_3(2)$, and $L_3(3)$ have the groups A_5, $L_3(2)$, and $L_3(3)$, respectively. Similarly, it can be shown that the groups of the 6-transposition quilts for A_6 and A_7 are the Schur covers $3.A_6$ and $3.A_7$, respectively, and that the $7 \cdot 42(3, 4^4, 5, 6^3)$ quilt for A_8 in Fig. 11.5.6 has a group of shape $2^7.(2 \times A_8)$. However, outside of such relatively small examples, no 6-transposition quilts have been found to have finite groups. In fact, we believe that their groups are almost always infinite; see Chap. 12.

12. Some results on the structure problem

In this chapter, we present some results related to the structure problem. Briefly summmarized, the evidence so far points strongly to the conclusion that a quilt/Norton system does not often strongly determine the (finite) group from which it came.

All quilts in this chapter are snug.

12.1 Finite quilts with infinite groups

The following examples show that snug quilts can be arbitrarily vague in specifying their groups. To be more precise, we exhibit finite quilts Q whose groups $G(Q)$ have arbitrarily large finite quotients, and therefore, arbitrarily large finite quotients with the same quilt. For all of these Q, we used the Reidemeister-Schreier facilities of GAP [77] to find successive commutator subgroups of a polyhedral presentation of $G(Q)$. The results are listed in Table 12.1.1.

All commutator notation used in Table 12.1.1 is explained in Sect. 2.1. Furthermore, in this section, all quilts are collapse-free, so we omit all arrow flows and dashes.

Table 12.1.1. Finite quilts with infinite groups

Q	Fig.	Quotient of $G = G(Q)$
$4{\cdot}24(4^6)$	12.1.1	$G/G^{(3)} \cong (4 \times \mathbf{Z}^4).(2^4 \times 4).4^2$
$5{\cdot}30(4^5, 5^2)$	12.1.2	$G/G^{(4)} \cong (8 \times \mathbf{Z}^{20}).2^8.5.2$
$6{\cdot}36(3^2, 5^6)$	12.1.3	$G/G^{(3)} \cong (3 \times \mathbf{Z}^{10}).3^4.5$
$8{\cdot}48(4^2, 5^8)$	12.1.4	$G/G^{(3)} \cong (2 \times \mathbf{Z}^{30}).4^4.5$
$10{\cdot}60(5^{12})$	12.1.5	$G/(G_{(2)})' \cong \mathbf{Z}^{20}.5.5^2$

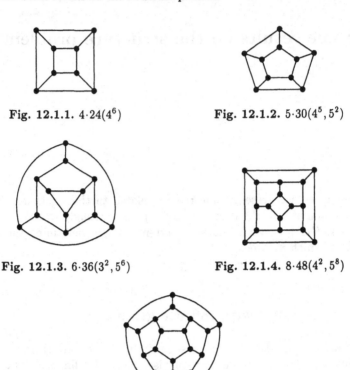

Fig. 12.1.1. $4 \cdot 24(4^6)$

Fig. 12.1.2. $5 \cdot 30(4^5, 5^2)$

Fig. 12.1.3. $6 \cdot 36(3^2, 5^6)$

Fig. 12.1.4. $8 \cdot 48(4^2, 5^8)$

Fig. 12.1.5. $10 \cdot 60(5^{12})$

12.2 Perfect quilts and perfect groups

If the group of a quilt is perfect, it cannot be analyzed using commutators, so we need more sophisticated methods to determine whether it is infinite. The quilts with perfect groups are called *perfect quilts*.

The easiest and most useful way of proving that a quilt is perfect comes from the following observation: Let Q be a quilt with group G. If Q has a patch whose stack elements have order m adjacent to a patch whose stack elements has order n, the order of G/G' must divide mn. For instance, this means that if Q has two adjacent unramified patches of order 5 and two adjacent unramified patches of order 6, then Q is perfect.

Another useful observation is that Q is perfect if and only if it covers no abelian quilts. We can then use the fact that the possible collapse in an abelian quilt is highly restricted (Thm. 5.4.2 and the preceding discussion) to show that Q is perfect. For instance, if Q has a collapsed vertex, it suffices to check whether Q covers the quilt for C_3, since that is the only abelian quilt with a collapsed vertex. Similarly, if Q has a collapsed edge, or if the modulus of Q is odd, it suffices to check if Q covers the quilt for C_2. Finally, if 12 does not divide the number of seams of Q, it suffices to check if Q covers either the quilt for C_2 or the quilt for C_3 (Thm. 6.1.14).

On the other hand, suppose we want to show that a quilt Q covers the quilt of an abelian group A. If we can reduce the arrow flows of Q mod 2, and label the patches of Q with elements of A, such that the appropriate relations hold at every vertex, Thm. 4.3.6 implies that Q covers the quilt of A. For example, consider the $30 \cdot 60(2, 3, 4^2, 5, 6^7)$ quilt for Co_3 shown in Fig. 11.2.18. Fig. 12.2.1 shows that this quilt covers the quilt of C_3, and is therefore not a perfect quilt. In fact, if G is the group of the quilt in Fig. 11.2.18, a calculation shows that $G/G' \cong 6$, and $G'/G^{(3)} \cong 2^{12}.5^2$, so G is quite far from being a perfect group.

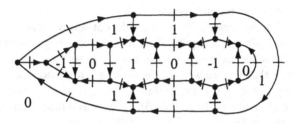

Fig. 12.2.1. Proof that quilt for Co_3 covers quilt of C_3

Some other examples of non-perfect quilts for perfect groups include the quilts for A_8 shown in Figs. 11.5.5–11.5.7, p. 145, the quilts for A_{12} shown in Figs. 11.5.8 and 11.5.9, p. 145, and the quilts for $U_4(2)$ shown in Figs. 11.6.3 and 11.6.5, pp. 147–148. These examples are not quite as dramatic as the example from Co_3, as for these other examples, if G is the quilt group, $G/G' \cong 2$, and $G' = G''$.

Fig. 12.2.2. Perfect quilt with infinite group

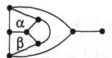

Fig. 12.2.3. Another perfect quilt with infinite group

Finally, let

$$G_1 = \langle \alpha, \beta \, | \, 1 = \alpha^4 = \beta^4 = (\alpha\beta)^3 = (\alpha^2\beta)^5 = (\alpha^{-1}\beta)^5 \rangle, \qquad (12.2.1)$$

$$G_2 = \langle \alpha, \beta \, | \, 1 = \alpha^4 = \beta^4 = (\alpha\beta)^3 = (\alpha^2\beta)^5 = (\alpha^{-1}\beta)^6 \rangle, \qquad (12.2.2)$$

be the groups of the perfect quilts in Figs. 12.2.2 and 12.2.3, respectively. W. Plesken and T. Schulz (private communication) have shown that G_1

(resp. G_2) is infinite, by finding a representation of G_1 (resp. G_2) onto a 21-dimensional (resp. 20-dimensional) space group, that is, an affine symmetry group of a 21-dimensional (resp. 20-dimensional) lattice. In fact, since the quilt in Fig. 12.2.2 (resp. 12.2.3) is also a quilt for $M_{21} \cong L_3(4)$ [44] (resp. M_{12}, see Sect. 12.3), this result gives two examples of a quilt for a finite simple group whose group is infinite. For more on the techniques used by Plesken and Schulz to find these representations, see Plesken and Souvignier [70].

12.3 Shadowing

When drawing quilts in this section, we omit all dashes, and we also omit all arrow flows except for patch inflow.

Now, since the methods of Sect. 12.1 do not apply to perfect quilts, one might still hope that perfect quilts determine their groups more closely than other quilts. However, except for a few cases, we suspect this is not so. One general problem that occurs with perfect quilts is a phenomenon that we call *shadowing*. Suppose α and β generate a group G_1, and α' and β' generate a group G_2. Let Q be the quilt of $T(\alpha, \beta)$. If there is a morphism from $T(\alpha, \beta)$ to $T(\alpha', \beta')$ sending (α, β) to (α', β'), it follows that

$$T(\alpha \times \alpha', \beta \times \beta') \cong T(\alpha, \beta). \tag{12.3.1}$$

In that case, if $\alpha \times \alpha'$ and $\beta \times \beta'$ generate $G_1 \times G_2$, then $G_1 \times G_2$ is a homomorphic image of $G(Q)$. This means that Q determines its group quite loosely, as it does not detect the projection from $G_1 \times G_2$ down to G_1.

For instance, we have the following example from A_7. Let

$$\begin{aligned}
\alpha &= (0\ 2\ 5\ 6\ 4), & \alpha' &= (0\ 2\ 5\ 3\ 4), \\
\beta &= (0\ 1\ 2\ 3\ 4\ 5\ 6), & \beta' &= (0\ 1\ 2\ 3\ 4\ 5\ 6).
\end{aligned} \tag{12.3.2}$$

Both $T(\alpha, \beta)$ and $T(\alpha', \beta')$ have the $12 \cdot 24(1_4, 3, 4^2, 5, 7)$ quilt shown in Fig. 12.3.1, with the same basepoint seam, which means that $T(\alpha \times \alpha', \beta \times \beta')$ has this quilt as well. Furthermore, $\alpha \times \alpha'$ and $\beta \times \beta'$ generate $A_7 \times A_7$, so as shown above, the group of the quilt in Fig. 12.3.1 has $A_7 \times A_7$ as a homomorphic image.

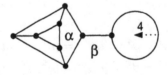

Fig. 12.3.1. Shadowing in A_7, $12 \cdot 24(1_4, 3, 4^2, 5, 7)$

We also have the following example (first published in Conway and Hsu [17]) from M_{12}. Using the projective numbering from Conway and Sloane [20, Chap. 10, §1], let

$$
\begin{aligned}
\alpha &= (0\ \infty)(1\ 5\ 8\ 9)(2\ 10)(3\ 4\ 7\ 6),\\
\alpha' &= (0\ 6\ 5\ 1)(2\ 9\ \infty\ 10)(3\ 8)(4\ 7),\\
\beta &= (0\ 6\ \infty\ 10)(1\ 5\ 3\ 7),\\
\beta' &= (0\ 9\ 6\ 8)(5\ 10\ 7\ \infty).
\end{aligned}
\tag{12.3.3}
$$

Both $T(\alpha,\beta)$ and $T(\alpha',\beta')$ have the $11\cdot 22(3,4^2,5,6)$ quilt shown in Fig. 12.2.3, p. 153, with the same basepoint seam. It is perhaps worth noting that this example was obtained from the following involution M(3)-systems, with $\alpha = ab$, $\beta = bc$, $\alpha' = a'b'$, and $\beta' = b'c'$:

$$
\begin{aligned}
a &= (2\ 10)(3\ 4)(5\ 9)(6\ 7) & a' &= (0\ \infty)(1\ 10)(2\ 5)(3\ 7)(4\ 8)(6\ 9)\\
b &= (0\ \infty)(1\ 5)(3\ 7)(8\ 9) & b' &= (0\ 10)(1\ 2)(3\ 4)(5\ 9)(6\ \infty)(7\ 8) \quad (12.3.4)\\
c &= (0\ 10)(1\ 3)(6\ 12)(8\ 9) & c' &= (0\ 7)(1\ 2)(3\ 4)(5\ 6)(8\ \infty)(9\ 10).
\end{aligned}
$$

The involutions a, b, and c are in conjugacy class $2B$ in M_{12} (in ATLAS [16] notation), whereas a', b', and c' are in class $2A$. In fact, as mentioned in Sect. 12.2, W. Plesken and T. Schulz have shown that the $11\cdot 22(3,4^2,5,6)$ quilt for M_{12} has an infinite group.

We also have the quilt Q_1 for $U_3(5)$ shown in Fig. 12.3.2. The reduction of Q_1 mod 5 is the quilt Q_2 shown in Fig. 12.3.3. It turns out that the group of Q_2 is the Schur central extension $(2^2 \times 3).M_{21}$, which means that $U_3(5) \times ((2^2 \times 3).M_{21})$ is a homomorphic image of the group of Q_1.

Fig. 12.3.2. Shadowing: $10\cdot 30(3_5, 4^3, 5^3)$

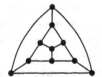

Fig. 12.3.3. Shadowing, $5\cdot 30(3, 4^3, 5^3)$

Finally, recall that the $15\cdot 90(4^3, 5^6, 6^8)$ quilt in Fig. 11.3.5, p. 140, is a quilt for Fi_{21}, Suz, and A_{12}, and the $11\cdot 66(3, 4^2, 5^5, 6^5)$ quilt in Fig. 11.4.8, p. 143, is a quilt for both Suz and A_{12}. The groups of these quilts are clearly quite large, and almost certainly infinite.

Question 12.3.1. Does shadowing mean anything? Is there some important sense in which the two copies of A_7, or the two copies of M_{12}, look alike, other than the fact that they have the same quilts? Is there some interesting way in which $U_3(5)$ resembles M_{21}, or Fi_{21}, A_{12}, and Suz all resemble one another?

We suspect that shadowing just means that quilts do not determine their groups very tightly. However, other explanations are certainly possible.

12.4 Monopoint Norton systems

The concept of *monopoint Norton system* arose during the early development of quilts. At that time, the action on conjugacy classes of pairs of elements of G was being considered, instead of the action of $\mathbf{B}_3$ on pairs of elements of G. Therefore, a pair whose Norton system was contained in a single orbit under conjugation was considered to have a trivial quilt. Such a Norton system is called a monopoint Norton system, and the only known prominent example was a Norton system for the Monster. It was therefore originally hoped that a group with such a Norton system might have special moonshine-related properties; in fact, it seemed possible that the Monster is the only group with a monopoint Norton system. However, as we shall see, this is not the case.

We looked for monopoint Norton systems by looking for *monopoint quilts*, that is, quilts whose standard Norton systems are monopoint. First, since Norton's action (4.2.1) commutes with conjugation, a monopoint quilt Q must represent a *normal* subgroup of $\mathbf{B}_3$. Therefore, since all of the patches of Q must have the same number of sides, one place to look for monopoint quilts is the quilts of genus 0 that look like regular polyhedra.

Because conjugation and Norton's action both commute with group homomorphisms, if Q is monopoint, any quilt Q covers is also monopoint. In particular, any monopoint quilt must be perfect, which rules out the dihedron $1{\cdot}6(2^3)$, the tetrahedron $2{\cdot}12(3^4)$, and the cubes $1{\cdot}24(4^6)$, $2{\cdot}24(4^6)$, and $4{\cdot}24(4^6)$. (The tetrahedron $1{\cdot}12(3^4)$ is not snug, as its group is trivial.)

The same argument holds for the dodecahedron quilts $2{\cdot}60(5^{12})$ and $10{\cdot}60(5^{12})$. On the other hand, the dodecahedron quilt $5{\cdot}60(5^{12})$ *is* perfect, and its group is $U_3(4)$. A calculation then shows that the standard Norton system for this quilt is monopoint, providing a counterexample to the above conjecture. As an aside, we note that $U_3(4)$ is a homomorphic image of the group of $10{\cdot}60(5^{12})$, which means that the latter group is *not* solvable, even though it has an infinite solvable quotient (see Table 12.1.1).

As for monopoint quilts with hexagonal patches (and therefore, genus 1), we have found a Norton system for $U_3(5)$ whose pairs have 3 orbits under conjugation in $U_3(5)$. The associated quilt is the genus 1 quilt from Exmp. 8.4.9. In fact, the 3 orbits of this Norton system fuse if we adjoin the action of an outer autmorphism of $U_3(5)$ of order 3, so in some sense, this example is almost a monopoint Norton system.

Independently, Norton [68] has also found a $7{\cdot}294(6^{49})$ monopoint quilt for the Held group, which is the nonabelian composition factor of the centralizer of an element of type $7B$ (in other words, type $7+$) in the Monster.

13. Further directions

In this chapter, we discuss other work on quilts, including some recent work, and some open questions. Other open questions can be found in Sect. 12.3.

13.1 Other work on quilts and Norton systems

As mentioned in the introduction (Sect. 1.2), the first published formal definitions of quilts and Norton systems were in Conway and Hsu [17]. Those definitions was revamped significantly in [43], which in turn formed the basis for much of this monograph. An enumeration of the 6-transposition quilts for the Mathieu groups can be found in [44], and an abstract of the material in Chaps. 3 and 4 (for B_3-quilts only) can be found in [45]. Note that in previous versions of this material, Norton systems were called *T-systems*.

Most other work directly related to quilts is due to Norton, beginning, of course, with their invention [62]. In [64], Norton also considers the problem of determining which (finite simple) groups are generated by n 3-transpositions. (See Chaps. 10 and 11, as well as Sect. 13.3.) Norton's more recent work on quilts principally stems from [65], which contains the following interesting observation. (See Conway [13] for more on the Monster algebra and its correspondence with certain elements of the Monster.)

Theorem 13.1.1 (Norton). *For $i = 1, 2, 3$, let a_i be a 6-transposition of the Monster, and let t_i be its corresponding vector in the Monster algebra. Then the Monster algebra triple product $(t_1 - 2, t_2 - 2, t_3 - 2)$ is the same for any triple in the monodromy system of (a_1, a_2, a_3).* □

This invariant of an $M(3)$-system of 6-transpositions in the Monster is called its *weight*.

In [66], Norton investigates several aspects of 6-transposition quilts arising from the Monster, including character-theoretic methods for enumerating them. Some new terminology is also introduced, as summarized in the following glossary.

net of (a, b, c) The quilt of the monodromy system of (a, b, c), for a, b, c in the 6-transposition class of the Monster.
netball A net of genus 0.

honeycomb A net of genus 1. (The name comes from the fact that such a net must be collapse-free, and all of its patches must have size 6.)

flag A seam.

face A patch.

order of a net The order of the monodromy stacking element cba of the monodromy system (a, b, c) (see Sect. 10.2).

faithful A net is said to be faithful if its order is equal to its modulus.

order of a face P The order of a stack element for P.

collapsed order of a face P The order of P.

n-gon A patch of order n.

exponent The exponent of a patch, edge, collapsed edge, or collapsed vertex is its *outflow*, normalized to be the smallest possible nonnegative integer. The exponent of an uncollapsed vertex is -1.

symmetric net A net that is a mirror-isomorphic to itself.

Subsequently, Norton has also determined all subgroups of the Monster that are generated by 6-transpositions [67] and enumerated all nets of order less than or equal to 7 [68].

Finally, since slightly different notation has been used to describe the 3-string braid group in Norton [66], in earlier versions of this material [44, 45], and in this monograph, we present another table translating among these notations (Table 13.1.1), taking presentation (2.4.4) as our standard presentation for $\mathbf{B}_3$. (Note that the matrices for s and t in [66] are incorrect.)

Table 13.1.1. Names for elements of $\mathbf{B}_3$ and $\mathbf{PSL}_2(\mathbf{Z})$

In $\mathbf{B}_3$	In $\mathbf{SL}_2(\mathbf{Z})$	In [44, 45]	In Norton [66]
Z	$\begin{pmatrix} -1 & 0 \\ 0 & -1 \end{pmatrix}$	Z	z^{-1}
V_1	$\begin{pmatrix} 1 & 1 \\ -1 & 0 \end{pmatrix}$	V	s^{-1}
V_2	$\begin{pmatrix} 0 & -1 \\ 1 & 0 \end{pmatrix}$	E^{-1}	t
L	$\begin{pmatrix} 1 & 1 \\ 0 & 1 \end{pmatrix}$	L	x^{-1}
R	$\begin{pmatrix} 1 & 0 \\ 1 & 1 \end{pmatrix}$	R	y

13.2 Norton's action and generalized moonshine

As mentioned in the introduction (Sect. 1.2), Norton [62] introduced quilts and Norton systems to investigate (holomorphic) functions $F : (M \times M) \times H^2 \to C$ such that

$$F((g, h)\pi, z) = F((g, h), \pi(z)) \tag{13.2.1}$$

for $\pi \in B_3$, where the action on the left-hand side of (13.2.1) is Norton's action, and the action on the right-hand side of (13.2.1) is the action of the image of π in $\mathbf{PSL_2(Z)}$ on H^2. Subsequently, the following has become known as Norton's *generalized moonshine* conjecture.

Conjecture 13.2.1 (Norton). *There exists an $F : (M \times M) \times H^2 \to C$ such that:*

1. *F satisfies (13.2.1) for commuting g, h; and*
2. *Roughly speaking, for fixed g and varying $h \in C_M(g)$, the functions $F((g, h), z)$ determine moonshine for $C_M(g)$. For example, if*

$$F((g, h), z) = \sum_{n \in Z} H_n(h)q^n, \tag{13.2.2}$$

then H_n is a character of $C_M(g)$.

See the introduction (Sect. 1.1) for a brief description of the original Monstrous Moonshine conjectures.

The generalized moonshine conjecture may be interpreted in terms of conformal field theory on an orbifold (see Mason [56] and Tuite [85]) and seems to be related to elliptic cohomology (see Hirzebruch, Berger and Jung [40]). A version of the conjecture has been proven for M_{24} (Dong and Mason [26]), and the conjecture itself has been proven in the case where $\langle g, h \rangle$ is cyclic (Dong, Li, and Mason [25]). For more on generalized moonshine from a mathematical physics point of view, see Tuite [85], which is also a good starting point for the related mathematical physics literature.

In the current context, we ask:

Question 13.2.2. Can Conj. 13.2.1 be further generalized to non-commuting values of g, h in $F((g, h), z)$ in some interesting way?

The most prominent obstacle to this question is that there is no obvious replacement for the property of being a character of $C_M(g)$. Perhaps there exists a suitable generalization based on the "modularity" property of (13.2.1).

13.3 Enumerating snug quilts

The snug quilts in Chaps. 9 and 11 are certainly not a complete listing of interesting snug quilts. The following are some of the more interesting possible enumerations.

Question 13.3.1. Enumerate the snug n-transposition quilts for $n = 4, 5$. Perhaps more generally (see below), enumerate the snug quilts whose patches have stack elements of order ≤ 5. Note that it follows from curvature results (Cor. 6.2.3) that there are only finitely many quilts whose patches have stack elements of order ≤ 5; in fact, it is not too hard to generate a list of all of them.

Are the two classes of quilts the same? In other words, is there a snug quilt whose patches have stack elements of order ≤ 5 that is not the quilt of a monodromy system of 5-transpositions?

Question 13.3.2. It is not hard to see that there are infinitely many different modular quilts whose patches have order ≤ 6; for instance, take an octahedron, barycentrically divide it arbitrarily many times, and take the dual. It follows *a fortiori* that there are infinitely many quilts whose patches have order ≤ 6. Are there infinitely many *snug* quilts whose patches have stack elements of order ≤ 6?

Question 13.3.3 (Norton). In [64], Norton constructs a group $G(n)$, called the *free 3-transposition group* of rank n, that contains a normal subgroup N such that $G(n)/N$ is a subdirect product of every group generated by n 3-transpositions. Is there a (tractable?) free m-transposition group of rank n for $m = 4, 5, 6$? How about rank 3?

Question 13.3.4 (Norton). Enumerate the quilts coming from triples of Monster 6-transpositions. See Norton [66] (described in Sect. 13.1) for possible methods, and [68] for the beginnings of such an enumeration.

13.4 Quilt groups

The results of Chaps. 8 and 12 lead to the following question.

Question 13.4.1. Do most snug quilts of modulus > 2 have infinite groups? Do only finitely many families of snug quilts of modulus > 2 have finite groups?

It may be better to restrict the scope of Ques. 13.4.1 to certain classes of quilts. Possible interesting classes include 6-transposition quilts, quilts whose patches have order ≤ 6, quilts of genus 0, and perfect quilts (Sect. 12.2).

Now initially, Ques. 13.4.1 may seem intractable, since there are a bewildering array of possibilities to consider. On the other hand, we note that recently, all finite groups of the form

$$(\ell, m \mid n, k) = \langle \alpha, \beta \mid 1 = \alpha^\ell = \beta^m = (\alpha\beta)^n = (\alpha\beta^{-1})^k \rangle \qquad (13.4.1)$$

have been classified (Edjvet and Thomas [30]). Since any quilt group is simultaneously a quotient of several such $(\ell, m \mid n, k)$ groups, it might be interesting to apply similar techniques to the study of quilt groups.

For a survey of related work and techniques, see Thomas [84]. See also Edjvet [28], Edjvet and Howie [29], Edjvet and Thomas [30], Holt and Plesken [41], Howie and Thomas [42], and Plesken and Souvignier [70]. (In fact, as we saw in Sect. 12.2, the methods of Plesken and Souvignier have already been used to show that several perfect quilt groups are infinite.)

We also note that quilt groups may be given the following perhaps more natural setting. Let F_2 be the free group on the generators α and β. We say that $g \in F_2$ is *primitive* if it is part of a generating pair for F_2. As shown in the Appendix to Conway and Hsu [17], every primitive element of F_2 is conjugate to an element of a pair in $N(\alpha, \beta)$. Call a group $G = \langle \alpha, \beta \rangle$ *primitive-periodic* if every primitive word in α and β has finite order in G. It follows from the patch theorem (Cor. 4.3.7) that if Q is a finite quilt, and α and β are the standard generators of $G(Q)$, then $G(Q)$ is primitive-periodic with respect to the generators α and β.

Question 13.4.2. Classify the primitive-periodic groups. For instance, can the finite primitive-periodic groups be placed into finitely many families? Note that this is not the same as Ques. 13.4.1, since the group of a quilt of genus greater than 0 is not isomorphic to a polyhedral presentation. It is not clear, however, whether this makes the question easier or harder.

For connections with other problems, including the restricted Burnside problem, see Zelmanov [87].

13.5 Variations on Norton's action and Norton systems

Since it seems that a Norton system for a group G does not usually provide very much information about G (Chaps. 8 and 12), it is natural to wonder if some variation on Norton's action and/or Norton systems provides more information, while still remaining tractable and natural.

Question 13.5.1. Can Norton's action be usefully generalized to n-tuples of group elements or $n + 1$-tuples of involutions for $n > 2$? For instance, there is a well-known surjective homomorphism from $\mathbf{B}_6$ to $\mathbf{Sp}_4(\mathbf{Z})$ (4×4 integer symplectic matrices) related to the Burau representation (Birman [5, 6]) that may be relevant. See also Norton [64].

Question 13.5.2. Is there some other way of "enlarging" Norton's action or Norton systems to make them provide more information? For instance, is there some way of combining power-mapping with Norton's action?

Question 13.5.3 (Conway). While quilts and the canonical patch labelling (Sect. 4.3) are well-suited to making pictures of Norton systems, they are not obviously as well-suited to making pictures of involution M(3)-systems. Is there some way of labelling seams with triples of involutions that makes it easier to draw pictures of involution M(3)-systems directly, without having to draw their reductions first? Is there some variation on the basic definitions of quilts that makes this possible? For instance, is the dual complex of a quilt (Fig. 6.1.1) more suitable?

13.6 Quilts and central extensions

Returning to the topic of subgroups of Seifert groups and central extensions (Sect. 7.2), let Σ be a Seifert group, let $\overline{\Sigma} = \Sigma/\langle Z \rangle$, let Γ be a normal subgroup of $\overline{\Sigma}$, and let $G = \overline{\Sigma}/\Gamma$.

Question 13.6.1. Suppose, for some nonnegative integer M and some subgroup $\Delta \leq \Sigma$, the following diagram of exact sequences commutes.

$$
\begin{array}{ccccccc}
 & 1 & & 1 & & & \\
 & \downarrow & & \downarrow & & & \\
1 \longrightarrow & \langle Z^M \rangle & \longrightarrow & \langle Z \rangle & & & \\
 & \downarrow & & \downarrow & & & \\
1 \longrightarrow & \Delta & \longrightarrow & \Sigma & & & \\
 & \downarrow & & \downarrow & & & \\
1 \longrightarrow & \Gamma & \longrightarrow & \overline{\Sigma} & \longrightarrow & G \longrightarrow 1 \\
 & \downarrow & & \downarrow & & \downarrow & \\
 & 1 & & 1 & & 1 &
\end{array}
\qquad (13.6.1)
$$

Under what conditions is Σ/Δ a central extension of G? In other words, when is Δ normal in Σ?

Let Q be the modular quilt of Γ in (13.6.1). Note that the automorphism group of Q is precisely G. Recall (Sect. 7.2) that if Q has genus 0, then Δ is always normal in Σ, but if Q has genus greater than 0, then Δ is often, perhaps even most of the time, *not* normal in Σ. Rephrasing Ques. 13.6.1 in these terms, we have:

Question 13.6.2. Let $\widetilde{Q}$ be the quilt of Δ. We have that Δ is normal in Σ if and only if the arrow flow of $\widetilde{Q}$ is invariant under the action of G, up to homology (exercise). Is there an *equivariant quilt theory* of such arrow flows? More importantly, given a modular quilt Q for $\Gamma \triangleleft \overline{\Sigma}$ and a compatible modulus M, is there a reasonable way of computing the (module of? group of?) quilts that are invariant under the action of G? This would give an interesting method for investigating the cyclic central extensions of quotients of triangle groups.

The following question should serve as a good test case for a proposed equivariant quilt theory.

Question 13.6.3. Recall that $L_2(7)$, the simple group of order 168, is the quotient of $(2\ 3\ \infty) \cong \mathbf{PSL}_2(\mathbf{Z})$ obtained by reducing $\mathbf{PSL}_2(\mathbf{Z})$ mod 7. In fact, since the resulting image of the element L of $(2\ 3\ \infty)$ in $L_2(7)$ has order

7, this exhibits $L_2(7)$ as a quotient of $(2\ 3\ 7)$. Let Q be the corresponding $(2\ 3\ 7)$-modular quilt, and let $\Sigma = \langle -\frac{1}{2}, \frac{1}{3}, \frac{1}{7} \rangle$. Now, since $e(\Sigma) = -\frac{1}{42}$ and Q has 168 seams, Σ is perfect (Thm. 2.3.6), and there exists a Σ-quilt with modular structure Q and modulus M if and only if M divides $\frac{168}{42} = 4$. In fact, since the genus of Q is 3, there exist $(\mathbf{Z}/4)^3$ such quilts. However, from the theory of the Schur multiplier (see, for instance, Aschbacher [2]), since the Schur multiplier of $L_2(7)$ is 2 (ATLAS [16]), none of the Σ-quilts with modular structure Q and modulus 4 represent normal subgroups of Σ (compare (7.2.1)). Can this be explained in terms of equivariant quilt theory?

13.7 Generalizations of quilts

It seems quite likely that our results on lifting subgroups of triangle groups to subgroups of Seifert groups can be extended to lifting subgroups of more general Fuchsian groups to subgroups of other fundamental groups of Seifert fibered 3-manifolds. For instance, consider groups of the form

$$\langle V_r\ (1 \leq i \leq r), L\,|\,1 = V_i^{m_i}\ (1 \leq i \leq r), 1 = L^n = V_1 \cdots V_r L \rangle. \qquad (13.7.1)$$

We may define modular quilts for such groups by starting with a modified "seam" with $2r$ outside dotted edges instead of $4 = 2(2)$ outside edges, as shown in Fig. 13.7.1. Modular quilt theory for such groups should essentially require no other changes; in fact, the corresponding quilt theory for groups of the form

$$\left\langle Z, V_i, L\ \middle|\ \begin{matrix} 1 = [Z, V_r] = [Z, L], \\ Z^{p_i} = V_i^{m_i},\ Z^r = L^n,\ V_1 \cdots V_r L = 1 \end{matrix} \right\rangle, \qquad (13.7.2)$$

where $1 \leq i \leq r$, should also extend with little difficulty.

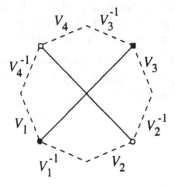

Fig. 13.7.1. A "seam" for presentation (13.7.1), $r = 4$

More generally, the fundamental group of an orientable Seifert fibered 3-manifold (see Seifert [79], or Orlik [69, Chap. 1, 5] and Scott [78, §1–3]) with orientable base space has the form

$$\Sigma = \Big\langle\ Z, A_j, B_j, V_i, L\ \big|\ 1 = [Z, A_j] = [Z, B_j] = [Z, V_r] = [Z, L],$$
$$Z^{p_i} = V_i^{m_i},\ Z^r = L^n, \tag{13.7.3}$$
$$V_1 \cdots V_r L = [A_1, B_1] \cdots [A_g, B_g]\ \Big\rangle,$$

where $1 \le j \le g$ and $1 \le i \le r$. Mod $\langle Z \rangle$, this becomes the Fuchsian group

$$\overline{\Sigma} = \left\langle A_j, B_j, V_i, L \ \middle|\ \begin{matrix} 1 = V_i^{m_i},\ 1 = L^n, \\ V_1 \cdots V_r L = [A_1, B_1] \cdots [A_g, B_g] \end{matrix} \right\rangle. \tag{13.7.4}$$

The parameter g represents the genus of the "base space" whose covers correspond with subgroups of $\overline{\Sigma}$.

Question 13.7.1. Do the results of Chaps. 3, 6, and 7 extend to groups of the form (13.7.3) and (13.7.4)? The principal new ingredient seems to be the consideration of base spaces of higher genus, but since most of the key theorems (for instance, Thm. 7.3.3) are stated purely in terms of the Euler number of Σ, perhaps our results generalize without modification.

Question 13.7.2. Is there a version of quilt theory that applies to *non-central* extensions of triangle groups? For instance, one might also look at surface groups extended by surface groups. Just as quilts can be interpreted as classifying complex line bundles over complex curves (Rem. 7.3.4), such generalized quilts might be relevant to the study of (singular) bundles of complex curves (or in other words, Riemann surfaces) with base space a complex curve. How about non-central extensions of general Fuchsian groups?

Question 13.7.3. Is there a version of quilt theory that can be used to investigate lifting subgroups of arbitrary finitely presented groups to their (central) extensions? On the one hand, we seem to have used the geometry of surfaces and triangle groups quite heavily, but on the other hand, perhaps this geometry may be replaced by more abstract homological ideas.

For some interesting work related to these last two questions, see Baik, Harlander, and Pride [4], who give a very general version of the results of Conway, Coxeter and Shepard [14].

A. Independent generators for modular subgroups

In this appendix, we consider the problem of obtaining certain "nice" systems of generators for subgroups of finite index in the classical modular group. Specifically, if the set $\{x_i\}$ generates a group Γ, we say that the x_i are *independent generators* if Γ is the free product of the cyclic groups $\langle x_i \rangle$. We demonstrate two methods of obtaining independent generators for a given modular subgroup, the first of which is simple and efficient (Sect. A.3), and the second of which produces particularly nice generators (Sect. A.4).

We remark that if the reader is willing to take Thm. A.3.2 and Cor. A.4.1 on faith, then Sect. A.4 does not depend on the rest of the appendix.

A.1 Notation for modular subgroups

Recall that since

$$\mathbf{PSL}_2(\mathbf{Z}) \cong (3, 2, \infty) = \langle V_1, V_2, L \,|\, 1 = V_1^3 = V_2^2 = V_1 V_2 L \rangle, \qquad \text{(A.1.1)}$$

conjugacy classes of subgroups of $\mathbf{PSL}_2(\mathbf{Z})$ are equivalent to transitive permutation representations of $(3, 2, \infty)$ (Thm. 3.1.1), which are in turn equivalent to modular quilts (Thm. 3.2.14). More precisely, by keeping track of basepoints, we obtain the following theorem.

Theorem A.1.1. *The following classes of objects are equivalent:*

1. *Subgroups Γ of $\mathbf{PSL}_2(\mathbf{Z})$;*
2. *Transitive permutation representations of $(3, 2, \infty)$ with a chosen basepoint, which by convention we call 1;*
3. *Modular quilts with a chosen distinguished (basepoint) seam.*

The correspondence between (1) and (2) (resp. (2) and (3)) is the correspondence of Theorem 3.1.1 (resp. 3.2.14). $\square$

Recall also that if we let $R = V_2 V_1$, then since

$$V_1 = RL^{-1}R, \qquad\qquad V_2 = R^{-1}L, \qquad \text{(A.1.2)}$$
$$L = V_2^{-1} V_1^{-1}, \qquad\qquad R = V_2 V_1, \qquad \text{(A.1.3)}$$

$\mathbf{PSL}_2(\mathbf{Z})$ is generated by any two of V_1, V_2, L, and R. In particular, specifying a permutation representation of $\mathbf{PSL}_2(\mathbf{Z})$ is equivalent to choosing transitive permutations V_1, V_2, L, and R satisfying (A.1.1)–(A.1.3), or in fact, transitive V_1 and V_2 such that $1 = V_1^3 = V_2^2$.

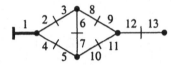

Fig. A.1.1. Basepointed modular quilt for Γ

Example A.1.2. Let Γ be the subgroup of index 13 in $\mathbf{PSL}_2(\mathbf{Z})$ whose basepointed modular quilt is shown in Fig. A.1.1. Then Γ is also described by any two of

$$
\begin{aligned}
V_1 &= (1\ 4\ 2)(3\ 6\ 8)(5\ 10\ 7)(9\ 11\ 12), \\
V_2 &= (2\ 3)(4\ 5)(6\ 7)(8\ 9)(10\ 11)(12\ 13), \\
L &= (1\ 2\ 8\ 12\ 13\ 11\ 5)(3\ 4\ 7)(6\ 10\ 9), \\
R &= (1\ 4\ 10\ 12\ 13\ 9\ 3)(2\ 6\ 5)(7\ 8\ 11).
\end{aligned}
\qquad (\text{A.1.4})
$$

A.2 Some combinatorial group theory

In this section, we describe two well-known pieces of combinatorial group theory, the first of which (Thm. A.2.9) is necessary for the sequel, and the other of which (Thm. A.2.12) provides a useful shortcut.

Now, while our first topic, the Reidemeister-Schreier method, has been described in many places, we give a complete account of it here, since we know of no suitable reference for our approach, which is modelled after lectures of Conway. For a more standard approach, see Magnus, Karass, and Solitar [54].

The Reidemeister-Schreier method is an algorithm that, given a presentation for a group G and a finite index subgroup $\Gamma \le G$, determines a presentation for Γ, while also keeping track of how the generators of Γ are expressed in terms of the generators of G. We begin our description of Reidemeister-Schreier with the following definition.

Definition A.2.1. Let G be a group generated by a set S, and let Γ be a subgroup of G. The *coset diagram* of $\Gamma \le G$ with respect to S is defined to be the directed graph X with the following nodes and edges.

1. X has a node associated with each coset Γa in G.
2. For every node Γa and every $s \in S$, X has a directed edge going from Γa to Γas, as shown in Fig. A.2.1. The edges associated with the generator s are called *s-edges*.

Note that X has a natural basepoint, namely, the coset Γ.

$$\Gamma a \quad \xrightarrow{\ s\ } \quad \Gamma as$$

Fig. A.2.1. Edge of a coset diagram

Exercise A.2.2. Show that if $G = (m_1, m_2, n)$ and $S = \{V_1, V_2\}$, then coset diagrams for subgroups of G with respect to S are equivalent to basepointed modular quilts. Indeed, coset diagrams are a more standard method for handling the material of Sect. 3.2.

Definition A.2.3. An *annotated coset diagram* of $\Gamma \leq G$ with respect to S is defined to be the coset diagram X of $\Gamma \leq G$ with respect to S, along with a choice of a spanning tree for X (see Thm. 2.6.6).

In Defns. A.2.4–A.2.7, let X be an annotated coset diagram of $\Gamma \leq G$ with respect to S, and let A be the set of edges of X not in the spanning tree. Eventually, we will show that A naturally defines an abstract generating set for Γ. We will also find corresponding defining relators for Γ and expressions for the subgroup generators A in terms of the original generators S.

Definition A.2.4. We define F_S to be the free group on S. We also let $\rho : F_S \to G$ be the corresponding natural map, and let $\tilde{\Gamma} = \rho^{-1}(\Gamma)$.

Definition A.2.5. For each node of X, we define a unique label $a \in F_S$ by the following rules:

1. The node of the coset Γ is labelled with 1.
2. If an s-edge of the spanning tree goes from a node with representative a to a node with representative b, then $b = as$ in F_S, as shown in Fig. A.2.2.

We denote the set of node labels of X by X_0. Note that because of the way we have defined our labels, for $a \in X_0$, a is the label of the node of the coset $\Gamma\rho(a)$. In other words, if a node is labelled a, then $\rho(a)$ is a member of the corresponding coset of Γ.

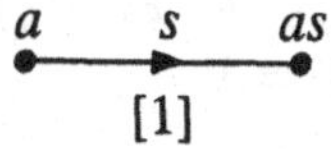

$[1]$

$[\alpha_i]$

Fig. A.2.2. Node labels on the spanning tree

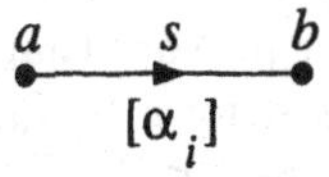

Fig. A.2.3. Edge labels outside the spanning tree

We next label each edge $\alpha_i \in A$ with $[\alpha_i]$, as shown in Fig. A.2.3, and each edge of the spanning tree with $[1]$, as shown in Fig. A.2.2.

Definition A.2.6. We define F_A to be the free group on A. By convention, we use Greek letters for elements of F_A and roman letters for elements of F_S.

We also define the homomorphism $w : F_A \to \tilde{\Gamma}$ by the rule that, for the s-edge $\alpha_i \in A$ travelling from the node labelled a to the node labelled b (Fig. A.2.3),

$$w(\alpha_i) = asb^{-1}, \qquad\qquad w(\alpha_i^{-1}) = bs^{-1}a^{-1}. \qquad (A.2.1)$$

Note that $asb^{-1} \in \tilde{\Gamma}$, since $\Gamma\rho(a)\rho(s) = \Gamma\rho(b)$.

As we will see, we may think of w as the function that rewrites the subgroup generators A in terms of the generators S of the full group. It may also be helpful to consider the commutative diagram (A.2.2), in which the horizontal arrows are subgroup inclusions.

$$
\begin{array}{ccc}
F_A & & \\
w \downarrow \uparrow \tau_1 & & \\
\tilde{\Gamma} & \longrightarrow & F_S \\
\rho \downarrow & & \rho \downarrow \\
\Gamma & \longrightarrow & G
\end{array}
\qquad (A.2.2)
$$

Note that (A.2.2) also contains a map τ_1, which we define as follows.

Definition A.2.7. For $a \in X_0$ and $x \in F_S$, we define $\tau(a, x) \in F_A$ by the following rules.

1. For an s-edge labelled $[\alpha]$ going from a node labelled a to a node labelled b, we define

$$\tau(a, s) = \alpha, \qquad\qquad \tau(b, s^{-1}) = \alpha^{-1}. \qquad (A.2.3)$$

   Note that α is either some $\alpha_i \in A$ or 1; see Figs. A.2.2 and A.2.3.
2. For $a, b \in X_0$ and $x, y \in F_S$ such that $\Gamma\rho(a)\rho(x) = \Gamma\rho(b)$, we define $\tau(a, xy) = \tau(a, x)\tau(b, y)$.

In other words, we define $\tau(a, x)$ by rule (1) and composition along the path x starting at a. Note that this is well-defined, since $\tau(a, xss^{-1}y) = \tau(a, xy)$. Note also that for $a \in X_0$, then $\tau(1, a) = \tau(a, a^{-1}) = 1$.

Finally, for $x \in \tilde{\Gamma}$, we define $\tau_1(x) = \tau(1, x)$.

Intuitively, we can think of $\tau(a, x)$ as representing the "error" that occurs in terms of coset representatives when we start at node a and travel through the path x. Alternately, we see from the following proposition that τ_1 is precisely the inverse of w.

Proposition A.2.8. *Let X be an annotated coset diagram of $\Gamma \leq G$ with respect to S, and let A be the set of edges of X not contained in the spanning tree. Then:*

1. *The map τ_1 is a homomorphism from $\widetilde{\Gamma}$ to F_A.*
2. *For $\alpha \in F_A$, $\tau_1(w(\alpha)) = \alpha$.*
3. *For $x \in \widetilde{\Gamma}$, $w(\tau_1(x)) = x$.*
4. *The map w is an isomorphism from F_A to $\widetilde{\Gamma}$.*

Proof. It is enough to prove assertions (1)–(3), since (4) follows easily. First, for $x, y \in \widetilde{\Gamma}$, $\Gamma\rho(x) = \Gamma$, so rule (2) of Defn. A.2.7 implies that

$$\tau_1(xy) = \tau(1, xy) = \tau(1, x)\tau(1, y) = \tau_1(x)\tau_1(y). \tag{A.2.4}$$

Assertion (1) follows.

Turning to assertion (2), it is enough to consider an arbitrary generator $\alpha_i \in A$ of F_A. However, for the generator α_i pictured in Fig. A.2.3, p. 167, we have

$$\tau_1(w(\alpha_i)) = \tau(1, asb^{-1}) = \tau(1, a)\tau(a, s)\tau(b, b^{-1}) = \tau(a, s) = \alpha_i. \tag{A.2.5}$$

Assertion (2) follows.

As for assertion (3), it again suffices to consider a generating set of $\widetilde{\Gamma}$. Now, for any s-edge going from a node labelled a to a node labelled b, we have

$$\begin{aligned} w(\tau(1, asb^{-1})) &= w(\tau(a, s)) = asb^{-1}, \\ w(\tau(1, bs^{-1}a^{-1})) &= w(\tau(b, s^{-1})) = bs^{-1}a^{-1}. \end{aligned} \tag{A.2.6}$$

It is therefore enough to show that $\widetilde{\Gamma}$ is generated by elements of the form asb^{-1} and $bs^{-1}a^{-1}$ for some s-edge going from a node labelled a to a node labelled b. So let $s = s_1^{\epsilon_1} \cdots s_n^{\epsilon_n}$ ($\epsilon_i = \pm 1$) be any element of $\widetilde{\Gamma}$. Let $a_0 = 1$, and for $i > 0$, let a_i be the label of the node of $\Gamma\rho(a_{i-1})\rho(s_i^{\epsilon_i})$. Now, since s is an element of $\widetilde{\Gamma}$, it must determine a path beginning and ending at the coset Γ. Therefore, the label a_n of the terminal node of this path is 1, and

$$s = s_1^{\epsilon_1} \cdots s_n^{\epsilon_n} = (a_0 s_1^{\epsilon_1} a_1^{-1}) \cdots (a_{n-1} s_n^{\epsilon_n} a_n). \tag{A.2.7}$$

The proposition follows. $\qquad\qquad\qquad\qquad\qquad\qquad\qquad\qquad\qquad\qquad\qquad$ $\square$

Theorem A.2.9 (Reidemeister-Schreier). *Let Γ be a subgroup of a group $G \cong \langle S \,|\, R \rangle$, where S is a generating set of G and R is a set of defining relators (words in the generators S that are equal to 1 in G). Let X be an annotated coset diagram of $\Gamma \leq G$ with respect to S, and let A be the set of edges of X not contained in the spanning tree. Then*

$$\Gamma \cong \langle A \,|\, \tau(a, r),\ r \in R \rangle. \tag{A.2.8}$$

In other words, if we take A as a generating set for Γ, then the consequences of applying each relator $r \in R$ at each node of X give a defining set of relators for Γ. In particular, if Γ has finite index in a finitely presented group G, then Γ is finitely presented.

Proof. First, since $\rho \circ w$ is a surjective map from F_A to Γ (assertion (4) of Prop. A.2.8), the image of A generates Γ. It is therefore enough to show that the kernel of $\rho \circ w$ is precisely the normal closure of the $\tau(a, r)$.

Now, on the one hand, for $r \in R$ and $a \in X_0$,

$$\tau(1, ara^{-1}) = \tau(1, a)\tau(a, r)\tau(a, a^{-1}) = \tau(a, r). \qquad (A.2.9)$$

Therefore, since $w(\tau(a, r)) = w(\tau(1, ara^{-1})) = ara^{-1}$ (assertion (3) of Prop. A.2.8), $\rho(w(\tau(a, r))) = 1$.

It therefore remains to show that the kernel of $\rho \circ w$ is generated by the conjugates of the $\tau(a, r)$. Now, by the definition of a presentation, if $\rho(w(\alpha_1^{\delta_1} \cdots \alpha_n^{\delta_n})) = 1$ for $\delta_i = \pm 1$, then

$$w(\alpha_1^{\delta_1} \cdots \alpha_n^{\delta_n}) = (x_1 r_1^{\epsilon_1} x_1^{-1}) \cdots (x_k r_k^{\epsilon_k} x_k^{-1}) \qquad (A.2.10)$$

in the free group F_S, for some $x_i \in F_S$, $r_i \in R$, and $\epsilon_i = \pm 1$. Since w is injective (assertion (4) of Prop. A.2.8), it is enough to show that for any $r \in R$ and $x \in F_S$, $xrx^{-1} = w(\sigma\tau(a, r)\sigma^{-1})$ for some $a \in X_0$ and $\sigma \in F_A$.

So for $r \in R$ and $x \in F_S$, let a be the label of the node of $\Gamma\rho(x)$. Since $xa^{-1} \in \widetilde{\Gamma}$, we may define $\sigma = \tau_1(xa^{-1})$. Then, since $w(\sigma) = w(\tau_1(xa^{-1})) = xa^{-1}$ (assertion (3) of Prop. A.2.8), we have

$$\begin{aligned} w(\sigma\tau(a, r)\sigma^{-1}) &= w(\sigma)w(\tau(a, r))w(\sigma)^{-1} \\ &= (xa^{-1})(ara^{-1})(xa^{-1})^{-1} \qquad (A.2.11) \\ &= xrx^{-1}. \end{aligned}$$

The theorem follows.   □

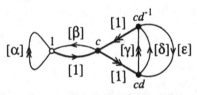

Fig. A.2.4. Annotated coset diagram for Exmp. A.2.10

Example A.2.10. Let $G = \langle c, d \, | \, c^8, (cd^{-1}c)^3 \rangle$, and let Γ be the subgroup of G with the annotated coset diagram shown in Fig. A.2.4. Note that in Fig. A.2.4, the basepoint is marked with a white vertex, the c-edges are marked with single arrows, the d-edges are marked with double arrows, and the chosen spanning tree is indicated by thick lines and the [1] edge labels.

We have

$$
\begin{aligned}
\tau(1, c^8) &= \beta^4, & \tau(c, c^8) &= \beta^4, \\
\tau(cd, c^8) &= (\delta\epsilon)^4 & \tau(cd^{-1}, c^8) &= (\epsilon\delta)^4 \\
\tau(1, (cd^{-1}c)^3) &= \epsilon\delta\gamma^{-1}\delta\epsilon\beta, & \tau(c, (cd^{-1}c)^3) &= (\beta\alpha^{-1})^3, \\
\tau(cd, (cd^{-1}c)^3) &= \delta\gamma^{-1}\delta\epsilon\beta\epsilon, & \tau(cd^{-1}, (cd^{-1}c)^3) &= \epsilon\beta\epsilon\delta\gamma^{-1}\delta,
\end{aligned}
\tag{A.2.12}
$$

so applying Reidemeister-Schreier and simplifying, we see that

$$
\Gamma \cong \left\langle \alpha, \beta, \gamma, \delta, \epsilon \,\middle|\, \beta^4, (\delta\epsilon)^4, (\beta\alpha^{-1})^3, \delta\gamma^{-1}\delta\epsilon\beta\epsilon \right\rangle.
\tag{A.2.13}
$$

Note that the node labels play no role in finding the presentation for Γ. In general, the node labels are only used to calculate the rewriting function w.

As for our other well-known piece of combinatorial group theory, we merely state the result we will need in the sequel.

Definition A.2.11. A group G is said to be *Hopfian* if every surjective homomorphism from G onto itself is an isomorphism.

Theorem A.2.12. *Finitely generated subgroups of* $\mathbf{PSL}_2(\mathbf{R})$ *are Hopfian.*

Proof. This follows because finitely generated subgroups of $\mathbf{PSL}_2(\mathbf{R})$ are residually finite (Lyndon and Schupp [53, Prop. 7.11]) and finitely generated residually finite groups are Hopfian (Lyndon and Schupp [53, Thm. 4.10]). $\quad\square$

A.3 Finite index modular subgroups

Let Γ be a subgroup of finite index n in $\mathbf{PSL}_2(\mathbf{Z})$. As mentioned in Sect. A.1, we may describe Γ by the corresponding transitive permutation representation Ω with basepoint 1; in fact, describing the actions of V_1 and V_2 on Ω is equivalent to describing the coset diagram of Γ with respect to $S = \{V_1, V_2\}$.

We would like to apply Reidemeister-Schreier to find a presentation for Γ and a rewriting function for its generators. Now, since the coset representatives of an annotated coset diagram play no role in determining a presentation for Γ, for the moment, we identify a coset $x \in \Omega$ with its representative, and we describe the edges of the coset diagram Ω by the pairs (x, s) with $x \in \Omega$, $s \in S$. Furthermore, since the function τ (Defn. A.2.7) is determined by composition along paths, to describe τ completely, we only need to describe the values of $\tau(x, s)$ for $s \in S$ and $x \in \Omega$. Finally, we observe that specifying a spanning tree for the coset diagram of Γ is equivalent to specifying which edges (x, s) have $\tau(x, s) = 1$. In short, to annotate Ω for the application of Reidemeister-Schreier, we need only to determine the values of $\tau(x, s)$.

We therefore come to the following procedure.

Definition A.3.1. Given a subgroup Γ of finite index n in $\mathbf{PSL}_2(\mathbf{Z})$, we have the following procedure for determining values of $\tau(x, s)$ for all $x \in \Omega$, $s \in S$.

1. For every V_1-orbit of size 3, pick some x in the orbit, and set $\tau(x, V_1) = \tau(xV_1, V_1) = 1$, as shown in Fig. A.3.1.
2. Mark the nodes $1, 1V_2, 1V_2^2 \in \Omega$ as path-connected to 1 within the spanning tree. (Note that we may have $1 = 1V_2 = 1V_2^2$.)
3. While there exists $x \in \Omega$ such that x is path-connected to 1 and xV_2 is not path-connected to 1:
   a) Set $\tau(x, V_2) = 1$.
   b) Mark $xV_2, xV_2V_1, xV_2V_1^2 \in \Omega$ as path-connected to 1.
4. For each $\tau(x, s)$ that has not been assigned a value, set $\tau(x, s) = \alpha$ for some new symbol α.

Fig. A.3.1. Labelling a V_1-orbit of size 3

Fig. A.3.2. Labelling a V_r-orbit of size 1 $(r = 1, 2)$

Theorem A.3.2. *Let Γ be a subgroup of index n of $\mathbf{PSL}_2(\mathbf{Z})$ whose modular quilt Q has genus g, n seams, p patches, e collapsed edges, and v collapsed vertices, and let Ω be the basepointed transitive permutation representation corresponding with Γ. Then*

1. *The procedure in Defn. A.3.1 yields a spanning tree for the coset diagram of $\Gamma \leq \mathbf{PSL}_2(\mathbf{Z})$ with respect to $S = \{V_1, V_2\}$.*
2. *Γ is isomorphic to the free product of e cyclic groups of order 2, v cyclic groups of order 3, and a free group of rank $2g + (p - 1)$.*

Proof. Let T be the edges (x, s) $(x \in \Omega, s \in S)$ such that $\tau(x, s) = 1$. From step (1) of Defn. A.3.1, it follows that every V_1-orbit of Ω is path-connected within T, and from step (3), it follows that every V_1-orbit is connected to node 1 within T. Therefore, T is connected, so to show that T is a spanning tree for Ω, we need only to show that T contains exactly $n - 1$ edges (Thm. 2.6.2). Examining Defn. A.3.1, we see that in step (1), we add 2 edges to T for each V_1-orbit of size 3, and in step (3), the number of edges we add to T is equal to the number of V_1-orbits minus 1. Therefore, since there are $\dfrac{n - v}{3}$ V_1-orbits of size 3 and v V_1-orbits of size 1 in Ω, the number of edges in T is

$$2\left(\frac{s-v}{3}\right) + \left(v + \frac{s-v}{3} - 1\right) = s - 1. \qquad (A.3.1)$$

Assertion (1) follows.

Fig. A.3.3. V_2-orbit with an edge   **Fig. A.3.4.** V_2-orbit outside T
in T

Applying Reidemeister-Schreier to our annotated coset diagram, we see that Γ has the following presentation.

1. For every V_1-orbit of size 3 (Fig. A.3.1), we have one generator, say, α. However, applying the relator V_1^3 at all three vertices of the orbit, we see that $\alpha = 1$.
2. For every V_1-orbit of size 1 (Fig. A.3.2), we have one generator, say, α, and applying the relator V_1^3, we see that $\alpha^3 = 1$.
3. Similarly, for every V_2 orbit of size 1 (Fig. A.3.2), we have a generator α such that $\alpha^2 = 1$.
4. For every V_2 orbit of size 2 with an edge in T (Fig. A.3.3), we have one generator α such that $\alpha = 1$.
5. For every V_2 orbit of size 2 with no edges in T (Fig. A.3.4), we have two generators α and β such that $\alpha\beta = 1$. In other words, we have one free generator α and one trivial generator $\alpha\beta$.

Now, since none of the elements of (1)–(5) interact in relations, Γ is the free product of the cyclic groups generated by them, so it remains only to count them. In fact, from (2), we get v cyclic groups of order 3 in the product, and from (3), we get e cyclic groups of order 2 in the product, so it remains only to count the rank of the free part. Examining (4) and (5), we see that the rank of the free part is equal to the number of V_2-orbits of size 2 that do not contain an edge of T. However, since the number of V_2-orbits of size 2 that contain an edge of T is equal to the number of edges added to T in step (3) of Defn. A.3.1, the rank of the free part is

$$\frac{n-e}{2} - \left(v + \frac{n-v}{3} - 1\right) = \frac{n}{6} - \frac{e}{2} - \frac{2v}{3} + 1. \qquad (A.3.2)$$

Then, recalling from Cor. 6.2.4 that

$$2 - 2g = -\frac{n}{6} + \frac{e}{2} + \frac{2v}{3} + p, \qquad (A.3.3)$$

we see that the rank of the free part is $2g - 2 + p + 1 = 2g + (p - 1)$. The theorem follows.   □

The proof of Theorem A.3.2 also yields the following consequence.

Corollary A.3.3. *By applying Reidemeister-Schreier to the annotated coset diagram from Defn. A.3.1, and simplifying as in the proof of Thm. A.3.2, we obtain a set of independent generators for Γ.*   □

In fact, Defn. A.3.1 essentially describes an algorithm for finding independent generators. It is also not hard to keep track of coset representatives while building the spanning tree in Defn. A.3.1, and thereby determine the values of the rewriting function w (Defn. A.2.6), as shown in the example below. We leave the details of the general case to the interested reader.

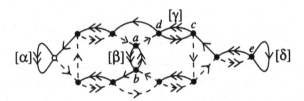

Fig. A.3.5. Annotated coset diagram for Exmp. A.3.4

Example A.3.4. Let Γ be the modular subgroup described in Exmp. A.1.2. Applying Defn. A.3.1 to the corresponding coset diagram, we obtain the annotated coset diagram shown in Fig. A.3.5, in which the basepoint is the white vertex, V_1-edges are indicated by single arrows, V_2-edges are indicated by double arrows, and the spanning tree is indicated by dashed lines. (Compare the modular quilt of Γ from Fig. A.1.1, p. 166.) By tracing paths through the spanning tree starting at the node labelled 1, we see that the indicated coset representatives are

$$a = V_1^2 V_2 V_1, \qquad b = V_1 V_2 V_1^2, \qquad c = (V_1 V_2)^2 V_1^{-1},$$
$$d = V_1^2 V_2 V_1^2, \qquad e = (V_1 V_2)^3. \tag{A.3.4}$$

Theorem A.3.2 then implies that

$$\alpha = V_2,$$
$$\beta = aV_2 b^{-1} = V_1^2 V_2 V_1 V_2 V_1^{-2} V_2^{-1} V_1^{-1} = V_1 (V_1 V_2)^3 V_1^{-1},$$
$$\gamma = cV_2 d^{-1} = (V_1 V_2)^2 V_1^{-1} V_2 V_1^{-2} V_2^{-1} V_1^{-2} = (V_1 V_2)^2 V_1^{-1} (V_2 V_1)^2, \tag{A.3.5}$$
$$\delta = eV_1 e^{-1} = (V_1 V_2)^3 V_1 (V_1 V_2)^{-3},$$

are a set of independent generators for Γ, of orders 2, ∞, ∞, and 3, respectively.

Remark A.3.5. We note that all of the material in this section may be generalized to subgroups of the free product of any two cyclic groups, and not just $C_2 * C_3$. For instance, essentially the same algorithm works for the Hecke groups $(2\ n\ \infty)$ $(n \geq 3)$.

The problem of finding independent generators for modular subgroups was posed by Rademacher [72], who used Reidemeister-Schreier to find independent generators for $\Gamma_0(p)$ (p prime). More recently, algorithmic methods for finding independent generators for congruence subgroups, relying on a combination of the Poincaré polygon theorem and arithmetic methods, have been developed by Kulkarni [49] and Lang, Lim, and Tan [50]. In particular, the reader may wish to compare the algorithm of Lang, Lim, and Tan [50, Sec. 3] that finds independent generators for a congruence subgroup Γ. See also Chan, Lang, Lim, and Tan [12].

A.4 Geometric independent generators

Throughout this section, let Γ be a subgroup of finite index s in $\mathbf{PSL}_2(\mathbf{Z})$ whose modular quilt Q has genus g, p patches, e collapsed edges, and v collapsed vertices.

Now, the appearance of $2g$ and $p-1$ in Thm. A.3.2 suggests that the free part of Γ should somehow be related to the patches and topology of Q. More specifically, recall that the nontrivial elements of $\mathbf{PSL}_2(\mathbf{Z})$ fall into three classes, namely, the *elliptic elements*, which are the conjugates of powers of V_1 and V_2, the *parabolic elements*, which are the conjugates of powers of L and R, and the *hyperbolic elements*, which are the other elements of $\mathbf{PSL}_2(\mathbf{Z})$. Since the parabolic elements are closely related to the patches of Q, one might hope that Γ is independently generated by e elliptic elements of order 2, v parabolic elements of order 3, $p-1$ parabolic elements, and $2g$ hyperbolic elements. We call such a generating set a set of *geometric independent generators* for Γ.

Now, the algorithm given in Sect. A.3 does not usually produce a geometric generating set; for instance, in Exmp. A.3.4, even though $g = 0$, by examining the trace of the matrix of γ, we see that γ cannot be conjugate to a power of V_1, V_2, L, or R, and is therefore hyperbolic. However, in this section, we show that a modified version of the results of Chap. 8 gives an algorithm for finding such a set of independent generators. The idea is that, just as relations in α and β allow us to close up the quilt of a Norton system in Chap. 8, knowing that certain elements are in Γ allows us to close up the modular quilt of Γ.

To begin with, we note the following consequence of Thm. A.3.2.

Corollary A.4.1. *Any set of e elements of order 2, v elements of order 3, and $2g + (p-1)$ other elements that generates Γ is a set of independent generators for Γ.*

Proof. Given such a generating set, from Thm. A.3.2, we can define a surjective homomorphism from Γ to itself by sending the standard generators of order 2 (resp. 3) to the new generators of order 2 (resp. 3), and so on. However, since Γ is Hopfian (Thm. A.2.12), this homomorphism must be an isomorphism, and the new generators must also be independent.   □

Proceeding to the algorithm itself, to simplify the exposition, we now assume that Q is collapse-free ($e = v = 0$).

Definition A.4.2. Let Q be a collapse-free basepointed modular quilt with p patches and genus g. Choose a spanning tree of polygons X for Q, and let A be the complementary cut to X.

Next, if $p > 1$, we label the polygons of X with elements of $\mathbf{PSL_2(Z)}$. Now, without loss of generality, we may assume that the basepoint seam of Q is in the interior of X, since we may move the basepoint by conjugating Γ, if necessary, and then conjugate back once we have found the desired generators. The following procedure therefore makes sense.

1. Label the polygons on the left and right of the basepoint (oriented in the manner of a basic edge) with L and R.
2. Label the remaining polyhedral labels using the following (recursive) rule: If two adjacent polygons P_1 and P_2 are labelled $\pi(1)$ and $\pi(2)$, with P_1 a left polygon and P_2 a right polygon, we label any previously un-labelled patch touching edge n of P_1 (resp. P_2) with $\pi(1)^n\pi(2)\pi(1)^{-n}$ (resp. $\pi(2)^n\pi(1)\pi(2)^{-n}$), as shown in Fig. A.4.1.

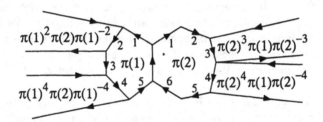

Fig. A.4.1. The conjugation labelling rule

Select a minimal (size $2g$) forced edge set (Defn. 8.4.2) A_0 for A, choose a polygon P_0 of X and call it the *cut polygon*, and let

$$S = \left\{ w(i)^{k(i)}, \varphi(a) \right\}, \tag{A.4.1}$$

where i runs over all polygons of X except P_0, $w(i)$ is the label of polygon i, $k(i)$ is the number of sides of polygon i, a runs over the forced edge set A_0, and $\varphi(a)$ is the edge-forcing element of a (Defn. 8.4.2). We call S a *quilt-selected* subset of Γ.

Note that all of the elements $w(i)$ in Defn. A.4.2 are conjugates of L and R, which means that the elements $w(i)^{k(i)}$ are parabolic. Furthermore, since the edge-forcing elements $\varphi(a)$ do not arise from a collapsed edge or vertex, or from "walking around" a patch, they must be hyperbolic.

Theorem A.4.3. *Let Γ be a finite index subgroup of $\mathbf{PSL}_2(\mathbf{Z})$ whose modular quilt Q is collapse-free with p patches and genus g. Any quilt-selected subset of Γ is a set of geometric independent generators for Γ.*

Proof. Let S be a quilt-selected subset of Γ. By construction, $\langle S \rangle \leq \Gamma$; so by the methods of Chap. 8, it suffices to show that the elements S allow us to recover Q. In the rest of the proof, we adopt the notation of Defn. A.4.2.

We first note that the parabolic elements $w(i)^{k(i)}$ allow us to reconstruct the spanning tree of polygons X, with one modification; namely, instead of recovering P_0, we recover P_0 cut open, as shown in Fig. A.4.2. Applying the edge-forcing elements $\varphi(a)$, we recover Q, cut open along the reduced complementary cut $A - A_0$ and the cut inside P_0.

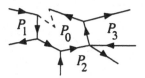

closes
to

Fig. A.4.2. A 5-sided cut polygon P_0, inside X

Fig. A.4.3. Closing the cut in the cut polygon

Now, since the cut inside P_0 just adds one edge and one vertex to the reduced complementary cut, the union of the reduced complementary cut and the cut inside P_0 is a tree A'. As in the proof of Thm. 8.3.1, we may then apply the trivalent rule to seal up all of A' except for the cut inside P_0. However, we then have the situation shown in Fig. A.4.3, so we may seal up this last cut as well.

It follows that S is a generating set for Γ containing $p - 1$ parabolic elements and $2g$ hyperbolic elements. We then see from Cor. A.4.1 that S is a set of independent generators, and the theorem follows. $\qquad\square$

Remark A.4.4. Note that the cut polygon argument in the proof of Thm. A.4.3 fails if we try to apply it to a B_3-quilt. The problem is that at the end of the proof, even though we may close up the cut in the cut polygon, as shown in Fig. A.4.3, we cannot then conclude that this last polygon is unramified, which means that we cannot apply our homology results (Cor. 6.1.8 and Thm. 6.1.10) to determine the modulus and arrow flows of the quilt.

Finally, we observe that for a modular subgroup Γ whose quilt has collapsed edges or vertices, we need only alter Defn. A.4.2 by replacing the spanning tree of polygons with a spanning tree of collapsed polygons (see Sect. 8.3) and adding the corresponding conjugates of V_1 or V_2 to the generating set S. We leave the details to the interested reader.

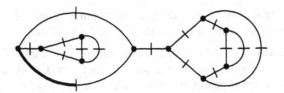

Fig. A.4.4. Basepointed modular quilt

Exercise A.4.5. Show that the subgroup $\Gamma \leq \mathbf{PSL}_2(\mathbf{Z})$ whose basepointed modular quilt is shown in Fig. A.4.4 is generated by the parabolic elements L^7, R^7, $L^4R^2L^{-4}$, $L^5R^3L^{-5}$, $R^2L^5R^{-2}$, and $R^3L^4R^{-3}$. (Compare presentation (8.2.2).)

References

1. M. Aschbacher, *The finite simple groups and their classification*, Yale math. monographs, vol. 7, Yale Univ. Press, New Haven, 1980.
2. _____, *Finite group theory*, Cambridge Univ. Press, 1986.
3. A. O. L. Atkin and H. P. F. Swinnerton-Dyer, *Modular forms on noncongruence subgroups*, Proc. Symp. Pure Math., Combinatorics (Providence) (T. S. Motzkin, ed.), vol. 19, AMS, 1971, pp. 1–26.
4. A. Baik, A. Harlander, and S. J. Pride, *The geometry of group extensions*, preprint, 1998.
5. J. S. Birman, *On Siegel's modular group*, Math. Ann. **191** (1971), 59–68.
6. _____, *Braids, links, and mapping class groups*, Princeton Univ. Press, Princeton, 1974.
7. R. E. Borcherds, *Monstrous moonshine and monstrous Lie superalgebras*, Invent. Math. **109** (1992), 405–444.
8. _____, *What is Moonshine?*, Documenta Mathematica, Proceedings of the International Congress of Mathematicians, Extra Vol. I (1998), 607–615, URL: http://www.mathematik.uni-bielefeld.de/documenta/.
9. R. Brauer and C. Sah (eds.), *Theory of finite groups*, W.A. Benjamin, Inc., 1969.
10. V. P. Burichenko, *On extensions of Coxeter groups*, Comm. Alg. **23** (1995), no. 5, 1867–1897.
11. R. Carter, *Simple groups of Lie type*, Wiley-Interscience, New York, 1972.
12. S.-P. Chan, M.-L. Lang, C.-H. Lim, and S.-P. Tan, *Special polygons for subgroups of the modular group and applications*, Internat. J. Math. **4** (1993), no. 1, 11–34.
13. J. H. Conway, *A simple construction for the Fischer-Griess monster group*, Invent. Math. **79** (1985), 513–540.
14. J. H. Conway, H. S. M. Coxeter, and C. G. Shephard, *The centre of a finitely generated group*, Tensor **25** (1972), 405–418, erratum in [15].
15. _____, *The centre of a finitely generated group (erratum)*, Tensor **26** (1972), 477.
16. J. H. Conway, R. T. Curtis, S. P. Norton, R. A. Parker, and R. A. Wilson, *ATLAS of finite groups*, Oxford Univ. Press, 1985.
17. J. H. Conway and T. Hsu, *Quilts and T-systems*, J. Alg. **174** (1995), 856–908.
18. J. H. Conway and S. P. Norton, *Monstrous moonshine*, Bull. Lon. Math. Soc. **11** (1979), 308–339.
19. J. H. Conway and A. D. Pritchard, *Hyperbolic reflections for the Bimonster and* $3Fi_{24}$, in Liebeck and Saxl [51], pp. 24–45.
20. J. H. Conway and N. J. A. Sloane, *Sphere packings, lattices, and groups*, 2nd ed., Springer-Verlag, New York, 1993.
21. H. S. M. Coxeter, *The abstract groups* $G^{m,n,p}$, Trans. AMS **45** (1939), 73–150.
22. H. S. M. Coxeter and W. O. J. Moser, *Generators and relations for discrete groups*, Springer-Verlag, Berlin, 1980.

23. C. J. Cummins and T. Gannon, *Modular equations and the genus zero property of moonshine functions*, Invent. Math. **129** (1997), no. 3, 413–443.

24. R. Diestel, *Graph theory*, Springer-Verlag, 1997.

25. C. Dong, H. Li, and G. Mason, *Modular invariance of trace functions in orbifold theory*, preprint q-alg/9703016, 1998.

26. C. Dong and G. Mason, *An orbifold theory of genus zero associated to the sporadic group M_{24}*, Comm. Math. Phys. **164** (1994), 87–104.

27. C. Dong and G. Mason (eds.), *Moonshine, the Monster, and related topics*, Contemp. Math., vol. 193, AMS, 1996.

28. M. Edjvet, *On certain quotients of the triangle groups*, J. Alg. **169** (1994), no. 2, 367–391.

29. M. Edjvet and J. Howie, *On the abstract groups $(3, n, p; 2)$*, JLMS **53** (1996), no. 2, 271–288.

30. M. Edjvet and R. M. Thomas, *The groups $(l, m|n, k)$*, J. Pure Appl. Alg. **114** (1997), no. 2, 175–208.

31. J. Ferrar and K. Harada (eds.), *The Monster and Lie algebras*, Ohio State Univ. Math. Res. Inst. Pubs., vol. 7, Berlin, de Gruyter, 1998.

32. R. D. Feuer, *Torsion-free subgroups of triangle groups*, Proc. AMS **30** (1971), 235–240.

33. P. Goddard, *The work of R. E. Borcherds*, Documenta Mathematica, Proceedings of the International Congress of Mathematicians, Extra Vol. I (1998), 99–108, URL: http://www.mathematik.uni-bielefeld.de/documenta/.

34. D. Gorenstein, *Finite groups*, 2nd ed., Chelsea, New York, 1980.

35. ———, *Finite simple groups: An introduction to their classification*, Plenum Press, New York, 1982.

36. D. Gorenstein, R. Lyons, and R. Solomon, *The classification of the finite simple groups*, vol. 40, Math. Surveys and Monographs, no. 1, AMS, 1994.

37. R. L. Griess, *Twelve sporadic groups*, Springer-Verlag, 1997.

38. M. Hall and D. Wales, *The simple group of order 604,800*, in Brauer and Sah [9], pp. 79–90.

39. A. E. Hatcher, *Algebraic topology*, Cornell Univ., 1999, to appear; also available electronically at http://math.cornell.edu/~hatcher/.

40. F. Hirzebruch, T. Berger, and R. Jung, *Manifolds and modular forms*, Friedr. Vieweg and Sohn, 1991.

41. D. F. Holt and W. Plesken, *A cohomological criterion for a finitely presented group to be infinite*, JLMS **45** (1992), no. 3, 469–480.

42. J. Howie and R. M. Thomas, *The groups $(2, 3, p; q)$, asphericity and a conjecture of coxeter*, J. Alg. **154** (1993), no. 2, 289–309.

43. T. Hsu, *Quilts, T-systems, and the combinatorics of Fuchsian groups*, Ph.D. thesis, Princeton Univ., 1994.

44. ———, *Some quilts for the Mathieu groups*, in Dong and Mason [27], pp. 113–122.

45. ———, *Quilts, the 3-string braid group, and braid actions on finite groups: an introduction*, in Ferrar and Harada [31].

46. A. A. Ivanov, *On the Buekenhout-Fischer geometry of the Monster*, in Dong and Mason [27], pp. 149–158.

47. G. A. Jones, *Congruence and non-congruence subgroups of the modular group: a survey*, Proceedings of Groups: St. Andrews (E. F. Robertson, ed.), Cambridge Univ. Press, 1985, pp. 223–234.

48. G. A. Jones and D. Singerman, *Maps, hypermaps and triangle groups*, in Schneps [74], pp. 115–145.

49. R. S. Kulkarni, *An arithmetic-geometric method in the study of the subgroups of the modular group*, Amer. J. Math. **113** (1991), no. 6, 1053–1133.

50. M.-L. Lang, C.-H. Lim, and S.-P. Tan, *Independent generators for congruence subgroups of Hecke groups*, Math. Z. **220** (1995), no. 4, 569–594.

51. M. W. Liebeck and J. Saxl (eds.), *Groups, combinatorics, and geometry, Durham, 1990*, Cambridge Univ. Press, 1992.

52. A. Lucchini, M. C. Tamburini, and J. S. Wilson, *Hurwitz groups of large rank*, To appear.

53. R. C. Lyndon and P. E. Schupp, *Combinatorial group theory*, Springer-Verlag, 1965.

54. W. Magnus, A. Karrass, and D. Solitar, *Combinatorial group theory*, 2nd ed., Dover Books, New York, 1976.

55. B. Maskit, *On Poincaré's theorem for fundamental polygons*, Adv. in Math. **7** (1971), 219–230.

56. G. Mason, *Remarks on moonshine and orbifolds*, in Liebeck and Saxl [51], pp. 108–120.

57. W. S. Massey, *Algebraic topology: An introduction*, GTM, no. 56, Springer-Verlag, 1977.

58. B. D. McKay, *nauty users guide, version 1.5*, Tech. Report TR-CS-90-02, Computer Science Dept., Australian National University, 1990.

59. M. H. Millington, *Subgroups of the classical modular group*, JLMS **1** (1969), 351–357.

60. J. R. Munkres, *Elements of algebraic topology*, Addison-Wesley, Redwood City, CA, 1984.

61. W. Neumann and F. Raymond, *Seifert manifolds, plumbing, μ-invariant and orientation reversing maps*, Algebraic and geometric topology: proceedings of a symposium held at Santa Barbara in honor of Raymond L. Wilder, July 25–29, 1977 (K. C. Millett, ed.), Lect. Notes Math., vol. 664, Springer-Verlag, 1978, pp. 163–196.

62. S. P. Norton, *Generalized moonshine*, The Arcata Conference on Representations of Finite Groups, Arcata, Calif., 1986 (P. Fong, ed.), Proc. Symp. Pure Math., vol. 47, AMS, 1987, pp. 208–209.

63. _____, *Constructing the Monster*, in Liebeck and Saxl [51], pp. 63–76.

64. _____, *Free transposition groups*, Comm. Alg. **24** (1996), 425–432.

65. _____, *The Monster algebra: some new formulae*, in Dong and Mason [27], pp. 433–441.

66. _____, *Netting the Monster*, in Ferrar and Harada [31].

67. _____, *Anatomy of the Monster: III*, preprint, 1999.

68. _____, *A string of nets*, preprint, 1999.

69. P. Orlik, *Seifert manifolds*, Lect. Notes Math., vol. 291, Springer-Verlag, Berlin, 1972.

70. W. Plesken and B. Souvignier, *Constructing representations of finite groups and applications to finitely presented groups*, J. Alg. **202** (1998), 690–703.

71. S. J. Pride and R. Stohr, *The (co)homology of aspherical Coxeter groups*, JLMS **42** (1990), no. 1, 49–63.

72. H. Rademacher, *Über die erzeugenden der kongruenzuntergruppen der modulgruppe*, Abh. Hamburg **7** (1929), 134–148.

73. L. Schneps, *Dessins d'enfants on the Riemann sphere*, in The Grothendieck Theory of Dessins d'Enfants [74], pp. 47–77.

74. L. Schneps (ed.), *The Grothendieck theory of dessins d'enfants*, Cambridge Univ. Press, 1994.

75. L. Schneps (ed.), *Geometric Galois actions II: Dessins d'enfants, mapping class groups and moduli*, LMS Lect. Notes, vol. 243, Cambridge Univ. Press, 1997.

76. B. Schoeneberg, *Elliptic modular functions*, Springer-Verlag, Berlin, 1974.

77. M. Schönert et al., *GAP: Groups, algorithms and programming*, Lehrstuhl D für Mathematik, RWTH Aachen, April 1992, Version 3.1.

78. P. Scott, *The geometries of 3-manifolds*, Bull. Lon. Math. Soc. **15** (1983), no. 5, 401–487.

79. H. Seifert, *Topology of 3-dimensional fibered spaces*, Seifert and Threlfall, A textbook of topology, Academic Press, 1980, pp. 359–437.

80. J.-P. Serre, *A course in arithmetic*, Springer-Verlag, 1973.

81. L. H. Soicher, *GRAPE: a system for computing with graphs and groups*, 1991 DIMACS Workshop on Groups and Computation, DIMACS Series in Discrete Mathematics and Theoretical Computer Science, to appear.

82. M. Spivak, *A comprehensive introduction to differential geometry*, vol. III, Publish or Perish, Inc., 1979.

83. M. Suzuki, *A simple group of order 448,345,497,600*, in Brauer and Sah [9], pp. 113–119.

84. R. M. Thomas, *Group presentations where the relators are proper powers*, Groups '93 Galway/St. Andrews: Proceedings of the International Conference held at University College, Galway, August 1–14, 1993 (C. M. Campbell et al., eds.), LMS Lect. Notes, vol. 211, Cambridge Univ. Press, 1995, pp. 549–560.

85. M. P. Tuite, *Generalised moonshine and abelian orbifold constructions*, in Dong and Mason [27], pp. 353–368.

86. H. Völklein, *Groups as Galois groups*, Cambridge Univ. Press, 1996.

87. E. Zelmanov, *On some open problems related to the restricted Burnside problem*, Recent Progress in Algebra (S. Korea, 1997) (S. G. Hahn, H. C. Myung, and E. Zelmanov, eds.), Contemp. Math., vol. 224, AMS, 1999, pp. 237–243.

Index

Printing: Weihert-Druck GmbH, Darmstadt
Binding: Buchbinderei Schäffer, Grünstadt

4. Lecture Notes are printed by photo-offset from the master-copy delivered in camera-ready form by the authors. Springer-Verlag provides technical instructions for the preparation of manuscripts. Macro packages in T_EX, L^AT_EX2e, L^AT_EX2.09 are available from Springer's web-pages at

http://www.springer.de/math/authors/b-tex.html.

Careful preparation of the manuscripts will help keep production time short and ensure satisfactory appearance of the finished book.

The actual production of a Lecture Notes volume takes approximately 12 weeks.

5. Authors receive a total of 50 free copies of their volume, but no royalties. They are entitled to a discount of 33.3 % on the price of Springer books purchase for their personal use, if ordering directly from Springer-Verlag.

Commitment to publish is made by letter of intent rather than by signing a formal contract. Springer-Verlag secures the copyright for each volume. Authors are free to reuse material contained in their LNM volumes in later publications: A brief written (or e-mail) request for formal permission is sufficient.

Addresses:

Professor F. Takens, Mathematisch Instituut,
Rijksuniversiteit Groningen, Postbus 800,
9700 AV Groningen, The Netherlands
E-mail: F.Takens@math.rug.nl

Professor B. Teissier
Université Paris 7
UFR de Mathématiques
Equipe Géométrie et Dynamique
Case 7012
2 place Jussieu
75251 Paris Cedex 05
E-mail: Teissier@ens.fr

Springer-Verlag, Mathematics Editorial, Tiergartenstr. 17,
D-69121 Heidelberg, Germany,
Tel.: *49 (6221) 487-701
Fax: *49 (6221) 487-355
E-mail: lnm@Springer.de